PANDEMONIUM

ALSO BY ANDREW NIKIFORUK

*School's Out: The Catastrophe in Public Education
and What We Can Do About It*

*The Fourth Horseman: A Short History of Plagues,
Scourges, and Emerging Viruses*

*Saboteurs:
Wiebo Ludwig's War Against Big Oil*

BIRD FLU, MAD COW DISEASE, AND OTHER BIOLOGICAL PLAGUES OF THE 21ST CENTURY

ANDREW NIKIFORUK

VIKING
CANADA

VIKING CANADA

Published by the Penguin Group

Penguin Group (Canada), 90 Eglinton Avenue East, Suite 700, Toronto, Ontario, Canada M4P 2Y3
(a division of Pearson Canada Inc.)

Penguin Group (USA) Inc., 375 Hudson Street, New York, New York 10014, U.S.A.
Penguin Books Ltd, 80 Strand, London WC2R 0RL, England
Penguin Ireland, 25 St Stephen's Green, Dublin 2, Ireland (a division of Penguin Books Ltd)
Penguin Group (Australia), 250 Camberwell Road, Camberwell, Victoria 3124, Australia
(a division of Pearson Australia Group Pty Ltd)
Penguin Books India Pvt Ltd, 11 Community Centre, Panchsheel Park, New Delhi – 110 017, India
Penguin Group (NZ), cnr Airborne and Rosedale Roads, Albany, Auckland 1310, New Zealand
(a division of Pearson New Zealand Ltd)
Penguin Books (South Africa) (Pty) Ltd, 24 Sturdee Avenue, Rosebank, Johannesburg 2196,
South Africa

Penguin Books Ltd, Registered Offices: 80 Strand, London WC2R 0RL, England

First published 2006

(RRD) 10 9 8 7 6 5 4 3 2 1

Manufactured in the U.S.A.

LIBRARY AND ARCHIVES CANADA CATALOGUING IN PUBLICATION

Nikiforuk, Andrew, 1955–
  Pandemonium : bird flu, mad cow disease & other biological plagues
of the 21st century / Andrew Nikiforuk.

ISBN-13: 978-0-670-04519-8
ISBN-10: 0-670-04519-5

1. Epidemics—Popular works. I. Title.

RA653.N55 2006   614.4   C2006-902566-5

Visit the Penguin Group (Canada) website at **www.penguin.ca**

Special and corporate bulk purchase rates available; please see
www.penguin.ca/corporatesales or call 1-800-399-6858, ext. 477 or 474

For Wendell Berry and David Schindler,
two men who refuse to compromise

And my boys, Aidan, Keegan, and Torin, good men-in-training

Contents

At *Pandaemonium,* the high Capital
Of Satan and his Peers: thir summons call'd
From every Band and squared Regiment
By place or choice the worthiest.
—John Milton

If disease is an expression of individual life under unfavorable
circumstance, then epidemics must be indicative of mass disturbances.
—Rudolf Virchow

Anyone who has studied science and talks to scientists notices
that we are in terrible danger now. Human beings,
past and present, have trashed the joint.
—Kurt Vonnegut

Introduction

On October 17, 2004, a Thai smuggler wrapped two small crested eagles from Tibet in cotton cloths. Then he placed each bird in a 60-centimeter (24-inch) wicker tube, making sure the raptors had room to breathe. With the tubes hidden in his hand luggage, the avian transporter boarded Eva Airways Flight BR0061 from Bangkok to Brussels via Vienna, along with 128 other jet-setters.

The smuggler was on a business trip. A Belgian falconer had ordered the birds for $17,000, and the avian entrepreneur had promised to make the delivery in Antwerp. But a random drug check at Zaventem airport in Brussels the next day uncovered the illicit cargo. Because bird flu had already killed 32 peasants and chicken handlers that year as well as millions of chickens and 83 tigers at Thai zoos, customs officials quarantined the birds and tested them. When both eagles proved positive for H5N1, authorities slaughtered 700 parrots and canaries in the quarantine facility. Twenty-five people who had come into contact with the feathery flu carriers were traced and given antiviral drugs. Authorities then tracked down the smuggler—they hadn't nabbed him earlier because importing diseased species is not a crime—and put him in an isolation ward at the Antwerp University hospital for four days.

The veterinarian who tested and killed the infected eagles developed conjunctivitis, a common flu symptom, just two days later. Doctors put his entire family on antiviral drugs. "We were very, very lucky," admitted Renee Snacken of Belgium's Scientific Institute of Public Health in Brussels. "It could have been a bomb for Europe."

This book is about biological bombs, or how globalization is unwittingly enabling a series of microbial invaders around the world. It explains how an enterprising Thai bird smuggler can pop up in Belgium and nearly ruin a continent's dinner in little more than a day; why Wyoming cowboys no longer have to go on African safaris to encounter West Nile fever; and how cholera, a native of Bangladesh, colonized most of the world's waters in just seven pandemics. But ultimately it's about the trade, travel, and eating habits of 6.5 billion people and all the biological hitchhikers that look to them for a free ride. It is a guide to instability, unpredictability, and the hidden microterrorists on our global doorstep.

Every day the Internet or the television news announces the arrival of another invader as the volume of global trade breaks another record. With one billion people on the move every year, we shouldn't be surprised that microbial traffic has also picked up. One day it's bird flu and the next it's methicillin-resistant *Staphylococcus aureus* (MRSA). Or white pine blister rust, cryptosporidium, or Lyme disease. Or Rift Valley fever, AIDS, or red tides. Or anthrax, mad cow disease, or the newest wheat fungus. Every microbe and its proverbial uncle seem to be emerging from the global trading network. "All of us, First World and Third, are paying the price of epidemiological globalization," notes Alfred Crosby, Jr., one of the world's most eminent disease historians. "We are homogenizing."

Although biological invaders usually start out as quiet strangers in paradise, they eventually go boom with unforeseen consequences. Whether colonizing port waters, park wildlife, hospital beds, or beef herds, their deadly work amounts to the same: the Wal-Martization of life. Invaders make the world a more uniform place where successful interlopers can sponsor more invasions. In fact, biological invasions are now driving approximately 50 percent of the world's 10 million species to extinction. Invaders simply erase local species and replace them with the globally fast and furious.

Consider for a moment the surreal progress of the brown snake. It arrived on the island of Guam on a U.S. military crate shortly after World War II. With nary a predator in sight, the invader slowly colonized the island, the same way bird flu colonized China's chicken factories in the 1990s. Nobody noticed the takeover until the island's native birds started to disappear. In the 1980s, beautiful natives such as the Rufus fantail and Micronesia broadbill vanished so quickly that scientists initially blamed their disappearance on a mysterious epidemic. After the snake wiped out eight species of forest birds (an estimated 300,000 birds in total), it started to dine indiscriminately on skinks, shrews, lizards, rats, chickens, and human babies. In recent years, the voracious and inquisitive invader has caused more than 100 power outages in Guam. Today the island boasts the highest snake density in the world: 4,600 per square kilometer (12,000 per square mile). Native plants that depend on birds to disperse their seeds are now losing ground.

Whether germs or snakes, biological invaders abide by a certain code that Malcolm Gladwell outlined in *The Tipping Point*. His witty book discussed the dynamics of social epidemics and showed how Hush Puppies and teenage smoking became such popular fads. The principles are simple: Successful products or ideas are contagious, little things lead to big sales, and change can often happen at lightning speed. Every winning product and fashion from iPods to oral sex has navigated these cultural pathways.

Biological invaders, like social epidemics, are either very contagious or very plastic. In other words, they can stick to a food-rich environment or readily adapt to any landscape. Bird flu found an unlimited viral replication machine in the world's burgeoning chicken factories, just as Lyme disease and its tick carriers found fertile soil in the world's expanding suburbs. Mad cow disease discovered the perfect pathway in global trade in animal feed protein, and MRSA has exploited overcrowding, filth, and patient mobility in the world's hospitals. Invaders thrive in the wake of human activity and trade.

Successful invaders also share certain pathological traits: They breed quickly and furiously, and they eat just about anything. (The gypsy moth feeds on more than 300 plant species, and MRSA can dine on almost any organ of the human body.) As determined opportunists, invaders remain ever alert to possibilities offered by man-made landscapes and climate change. Like thugs probing a corner grocery store, they check out the scene, mix with the locals, bide their time, and then burst in and empty the till.

Once invaders start to reproduce in exponential numbers, they can alter entire ecosystems, food chains, and water systems—and even the fate of human empires. In the process, they doggedly help other invaders join the new global zoo. Species juggling produces a few winners and many losers. The global winners—like the world's ultimate invader, *Homo sapiens*—diminish diversity, decrease disease resistance, replace local biology with global tyrants, and generally homogenize all life. They make the world either less interesting or more dangerous.

Invaders prove that global trade is not just an innocent exchange of goods and gadgets but *trade in every living thing.* Each economic enterprise provokes a commensurate biological transaction. Who would have thought that the explosive growth of chicken factories and trade in southeast China would transform a common virus into a global killer and stir up worldwide chaos? But it did. In one dramatic year alone, 2003, bird flu irrevocably changed the future of chickens and wild birds.

Invaders tend to make colorful social critics. Every day, globe-trotting fungi or bacteria determinedly expose a number of glaring insecurities in the way we live. International trade and reckless governance have made mad cow disease a universal citizen. The ease of modern travel (and the inclusive eating habits of Guangdong province) allowed SARS, a rather lazy virus, to travel abroad and illuminate the sorry state of hospital infectious-disease control. (Hospitals are among the world's most invaded institutions.) The troubling proliferation of 600 livestock diseases (from avian flu to foot-and-mouth disease) in the last two decades puts into question the wisdom of the "livestock revolution" and the bigger-is-better fad in agriculture. Legal and illegal science labs can now make engineered germs and weaponized microbes as fast as a livestock factory. The great ecologist Charles Elton observed nearly 50 years ago, "We are living in a period of the world's history when the mingling of thousands of kinds of organisms from different parts of the world is setting up terrific dislocations in nature." He predicted that the mayhem would become a constant stream of "unforeseen emergencies."

Our relentless mingling of pests, weeds, and germs, abetted by worldwide trade, creates weekly emergencies on every front. Every reader of this book will meet a trade-promoted invader someday, somewhere. The intruder might be an economic saboteur or a global killer. It might be as ambitious as H5N1 or as costly as SARS ($50 billion); as contagious as foot-and-mouth disease or as lethal as MRSA; as devastating as potato blight or as diabolical as anthrax. Rudolf Virchow, the distinguished 19th-century pathologist, once lamented that humanity's greatest curse was that "it can learn to tolerate even the most horrible situations by habituation."

The more insecure global invaders make our neighborhoods, the more we seem to grow accustomed to the mayhem. We have become grand traders in pandemonium.

ONE

Avian Flu:
Barbarians in the Coop

"Without chicken, there will not be a banquet."
—Chinese Proverb

For an ancient and yet nasty poultry virus, H5N1 has created a lot of pandemonium. First came the televised bird holocausts that featured solemn men dressed in ghostly killing gear. Reports of dying or dead peasants soon followed. Politicians then blamed the viral trade killer on wild migratory birds and ordered the draining of wetlands. After white coats at the World Health Organization in Geneva added up the dead (nearly 100 poor people), virologists started to fret about another global influenza pandemic like the viral blizzard of 1918 that dispatched 50 million people. Meanwhile Chinese authorities did what the modern Chinese do best: They issued a flock of denials about the pandemic's authentic birthplace, Guangdong.

An eerie Hitchcockian spell soon infected Internet blogs, public health officials, and poultry lovers everywhere. The hullabaloo trailed H5N1 wherever it popped up. In Egypt fearful peasants culled their chicken flocks, whether infected or not, and dumped their primary source of protein into the Nile. When dead birds fell from the sky in northern Iraq, people ran away as though nearly struck by God. The wary German government dispatched Tornado reconnaissance jets to locate hundreds of swans felled by flu along the shores of the Baltic Sea. The fastidious Dutch declared an emergency and moved all their poultry indoors. The French, whose proud national symbol remains a rooster, suffered an identity crisis. The Hong Kong authorities warned bird-loving citizens not to kiss their feathered pets. And governments everywhere started to publish pandemic preparation plans by the dozen.

But before avian influenza set the global village on fire with a rare gallinaceous fever, Mark Dekich met and did battle with the invader in a crowded Saudi Arabian chicken factory. The year was 1998. At the time Dekich's exploits in the oil kingdom didn't get much notice. In fact, Dekich went to criminal lengths to keep them quiet. The global poultry industry doesn't really like talking about its factory-made infections. But Dekich's exceptional story partly explains how a plain avian virus became a feared global menace and eventually worked up a hurricane of human hysteria.

Eight years ago Dekich held a good job at Fakieh Poultry Farms, one of the biggest chicken producers in the oil-rich kingdom. When not taking care of Fakieh's birds with mass vaccinations and medicated feeds, the Georgia-born bird doctor spoke at symposiums on factory chicken diseases and even testified for the U.S. Department of Agriculture on key livestock matters. As a longtime poultry expert, Dekich knew that "fowl plague" comes in two forms: mild maimers that clog up chicken lungs, and strains of "highly pathogenic avian influenza" that kill. Any maimer can morph into a killer and clean out a factory of 40,000 birds with 100 percent mortality in hours.

A pathogenic strain works with a certain demonic precision: It simply overwhelms meat birds the same way Ebola melts Africans. The invader commandeers every organ and tissue of a bird, from the top of its comb to the tips of its ammonia-burned feet. The broilers bleed from their eyes, beaks, and anuses. One moment they are pecking away; the next they fall over dead. In the end they just "melt out," resembling rubber caricatures of themselves. An outbreak not only eats up profits and creates unsightly piles of dead animals but generally shuts down a bird factory for weeks.

Like most people working in the global chicken trade, Dekich had nervously watched a viral outbreak crash and burn hundreds of thousands of birds in Hong Kong's high-density poultry farms and live markets in the spring and fall of 1997. The sudden plague wiped out between 70 and 100 percent of the ducks, chickens, and quails it encountered. Then a three-year-old boy played with a chick at a day-care center and came down with an unusual bout of the flu. The disease shut down his lungs, kidneys, and liver. He died six days later in a coma. When virologists identified this viral citizen as H5N1, they started to pace at night: The strain had never assaulted or outgunned the human immune system before. H5N1 was supposed to be a stable avian predator, not a pioneering human one.

Six months later the deadly virus emerged again after weeks of wet weather. It immediately went on another chicken-killing spree. After five more children and adults got caught in the viral crossfire and died of overwhelmed immune systems, Hong Kong's public health authorities took draconian measures. They sent out four-person teams to kill 1.6 million birds in the city's farms and "wet" markets, the bustling clusters of stalls where vendors of live birds and fish clean the streets by pouring buckets of water down them. The avian assassination squads broke necks, cut throats, and gassed birds in bags. When the masked killers came home at night, they did not kiss their wives or touch their children.

As Hong Kong battled the outbreak, Dekich penned a prescient paper for the journal *Poultry Science*. After 30 years of breeding "big, economical and fast growing chickens," wrote Dekich, the profitable commercial broiler business faced a number of health risks, including unreliable vaccines and "increased poultry house density." Giving chickens less space than a page of typing paper to live on just heightened the "potential for spread of endemic diseases through large naïve populations of birds." A "naïve" bird is an immunocompromised animal with no natural disease resistance.

Shortly after penning the warning, Dekich learned that some of his broilers at Fakieh's factories were keeling over from a mild strain of avian influenza. It wasn't H5N1 but H9N2, a weaker and rapidly evolving cousin that had started to rattle chicken cages in Asia, Europe, and the Middle East in the 1990s. H9N2 (every avian virus has what Mike Davis, author of *The Monster at Our Door*, quaintly calls a "genetic license-plate number") sickened factory chickens with lung problems and outright killed birds burdened by other infections. Because virologists then regarded H9N2 as one of the world's most abundant bird viruses, they placed it "high on the list with pandemic potential" for humans. Given the growing hysteria about H5N1 in Hong Kong, Dekich knew that any public revelation of the outbreak would result in serious investigations, trade restrictions, and multimillion-dollar losses. The same invader had already raised hell in Iran's chicken flocks.

So Dekich did what every major factory farmer and broiler-friendly government has done since avian flu first invaded the global chicken coop: He kept the barbarian under wraps. "With the world health publicity about avian influenza, it was a priority of the company to keep confidentiality. The human epidemiological consequences can be severe if this H9N2 virus is allowed to continue to replicate and re-sort its genome (genes). Fakieh Poultry Farms must go avian influenza negative and protect that negative status," he later wrote in one company memo.

After consulting with a well-known poultry specialist at the University of Delaware about his fluish flocks, Dekich arranged for a struggling biotech firm, Maine Biological Laboratories, to secretly produce a vaccine. Dekich then smuggled a sample of the chicken killer into the United States. MBL mislabeled $900,000 worth of its product as a vaccine for Newcastle disease (another viral menace) in order to evade detection by U.S. authorities.

Until Dekich and six employees of MBL pleaded guilty seven years later to mail fraud, virus smuggling, or making false statements to the government (among other crimes), no one knew much about the mail traffic in poultry killers or Saudi Arabia's avian flu outbreak in 1998. Upon sentencing Dekich and his accomplices in July 2005, U.S. District Judge John A. Woodcock, Jr., called the corruption "insidious, enticing and aggressive." The judge's verdict could easily have been applied to the global spread of avian flu.

Bird flu is now poised to sweep through humans like a viral tsunami. Most members of the world's unhappy flu-watching fraternity now debate only the severity of the next pandemic. Will the virus heat up and kill 1 in 50 people, setting off an economic collapse, or cool down and dispatch 1 in 1,000 people, setting off a global panic attack? Or will it simply fade away, leaving citizens with nothing more than a bad case of H5N1 fatigue and governments with expensive stockpiles of antiviral drugs, like Tamiflu?

Whatever its future course, H5N1 has already changed the world forever. It has done to chickens what AIDS did to sex, and it has visited upon Asia's poor the same kind of calamity Katrina dropped on the indigent in New Orleans. But that's just the beginning. The invasive opportunist has endangered the world's number two source of protein, accelerated species extinction among wild birds, impoverished more than 100 million rural families, demonized backyard chicken keepers, cost Asian taxpayers $20 billion, and added another layer of vulnerability to modern life. Like it or not, H5N1 has become another global mischief maker with a universal passport. Earl Brown, a University of Ottawa virologist, believes more viral mayhem is on the way: "Welcome to the beginning of the new normal."

T he Great Chicken Pandemic, which has buried more than 200 million winged creatures, is another unadvertised by-product of globalization.

Its source, pure and simple, is our gluttonous appetite for cheap, industrially produced meat. Crowded bird factories, rampant bird smuggling, bad vaccines, and duplicitous governments have all played a role in fouling the proverbial nest. Medical professionals may not like to admit it, but avian flu is a fairly predictable man-made plague, or what scientists cryptically call a "deliberately emerging microbe." Even the UN Food and Agriculture Organization (FAO) has repeatedly concluded that avian flu owes its global reach to "the intensification and concentration of livestock production in areas of high density human populations." H5N1, in other words, is a factory-acquired infection.

It is also a bona fide biological invader and a member of the fast-and-furious club. Anthony McMichael, a well-known Australian public health expert, has long argued that the roller-coaster pace of global trade and travel selects for troublemakers by disrupting their natural homes or ecosystems. As a result the world's r-species, true die-hard globalists such as fungi, viruses, and bacteria, are once again looking for new homes. Unlike k-species (apes and humans), the r-species reproduce quickly, do little parenting, and excel at spreading their numerous offspring. Trade, travel, city making, and chicken factories smile on "these small opportunistic species," says McMichael. Biologists just call them invaders.

Avian flu has been waiting for an opportunity to go global for a long time. This ancient virus has dwelled harmlessly in the guts of wild ducks and other migratory waders for millions of years. There it hijacks a duck's intestinal cells to make more evolutionary anarchists and gene spreaders. Whenever the duck defecates, the viral freeloaders enter ponds and lakes where they can survive happily for months at a time. It's telling that 10 billion virus particles can be found in a single liter or quart of water. In the scheme of things, God probably made ducks to breed and transport viruses.

Wild fowl visit ponds inhabited by domestic poultry. Here wild ducks share their viral cargo, much the way careless suburbanites share gastrointestinal trouble in hot tubs or ecotourists share measles with apes. Just a gram (0.03 ounce) of infected duck feces can infect a million chickens. Depending on the strain of influenza that the wild fowl transmit to them, the domestic birds can gain immunity, start sneezing, or fall down dead.

As the world's oldest pathogenic entrepreneurs, viruses are always looking for opportunities to expand their domain. This explains why influenza now comes in four major and often troublesome tribes: A primarily dwells in wild fowl but occasionally makes forays into several mammals, including horses, humans, and pigs; B, a tamer tribe, mostly dogs humans but has been found in seals; C can produce mild colds in humans and pigs; and Thogoto remains largely a tick-borne virus in Asia and Africa. All the tribes can occasionally swap genes and evolve in unpredictable ways, but only A and B can author a pandemic that turns human lungs and brains into a bloody mess.

The A tribe can stir up pandemic trouble in as many as 154 genetic combinations or subtypes. Each tribal subtype, in turn, can spawn hundreds of different strains. (Like fashion models, influenza viruses are always changing shape.) H5N1, for example, has mutated as many as 20 times already. The subtypes can be told apart by the position of two creative proteins on the surface of the avian virus: hemagglutinin (HA) and neuraminidase (NA). Their arrangement helps to shape a virus's ability to break and enter into a cell. All of these subtypes call birds home, but certain ones—H1, H2, and H3—stick pretty well to human lungs while others, such as H7, prefer horses. Until H5N1 popped out of Asia, scientists generally thought that no member of the A tribe could make a bad case of the flu without first swapping genes with other tribal groups or at the very least passing first through pigs, the world's best viral mixing medium. But H5N1 surprised them.

Like most RNA viruses, the avian clan tend to be sloppy and capricious reproducers. During "viral sex" they make all sorts of rude copies and in the process create innumerable mutants. A heavy load of flu virus (say, 50,000 particles) can make up to 50,000 mutants. In a chicken factory, a strain of avian flu will quickly evolve from a benign hitchhiker into a nasty killer. Crowds encourage the selection of deadlier and stickier forms of influenza, explains Earl Brown. "The virus responds to its environment, and the highest yielding virus on the barn floor wins." The winner is usually a formidable chicken killer.

Italians first described an outbreak of the "fowl plague" in 1897, but it wasn't until 1955 that scientists identified the culprit as a member of the avian influenza tribe. Although fowl plague killed off a fair number of domestic birds early in the century, it didn't really gain strength until we foolishly separated the chicken from viral evolution in wild birds.

That separation occurred when American ingenuity malled the chicken. After World War II the age-old practice of rearing a couple of feathered friends in the backyard got dumped for the four horsemen of the flu: density, efficiency, drugs, and big money. The principles of factory farming were simple: Take a sturdy and desirable stock, breed out qualities such as disease hardiness, and breed in a genetic disposition to grow fast on less feed. Then crowd the efficient meat producers into cages by the thousands. Add indoor lighting and air conditioning. A small number of companies supplied all the chicks and feed to "keepers," contractors who fattened the birds for automated slaughtering plants. By 1965 one man alone could operate a factory producing 40,000 chicks a day. Today one-fifth of the world's 100 billion birds are genetically uniform broiler chickens.

The industrialization of the once noble chicken had several effects: It made eggs and chicken meat much cheaper, displaced thousands of farmers in North America, and generated a series of appalling chicken epidemics. First came coccidiosis, an intestinal parasite that preyed on young chicks awash in too much of their own waste. Next appeared Newcastle disease. This novel virus leaped out of imported Hong Kong pheasants into California's poultry flocks in the 1950s. Since then Newcastle disease, an agent almost as infectious as avian flu, has developed a taste for factory birds around the world. An infamous 1972 outbreak resulted in the "stamping out" of 9 million chickens in California alone at a cost of $57 million.

Page Smith, an American journalist and chicken historian, found that poultry holocaust deeply disturbing: "Chickens are not people, but perhaps the destruction in our age of millions of human beings who were thought to carry a kind of racial virus in their genes has inured us to the horror of killing so many living creatures and left us equally indifferent to the strange developments which make such a solution seem inevitable." Smith had no idea that gallicide would soon become a regular staple of the global poultry trade.

To stamp out one pathogenic invader after another, the chicken business resorted to the usual quick fixes. It doled out antibiotics, vaccines, and medicated feeds in an effort to keep stressed-out, immune-blasted birds upright. (Most broilers are now so breast-heavy that many are lame or prone to heart attacks.) After desperate Hollywood housewives, broiler chickens are probably the most drugged denizens on the planet. Whenever a virus or bacterium gets into a factory and threatens feed schedules or weight gain, the "units" are routinely rounded up and rubbed out. Frank Perdue, the American chicken magnate, wasn't kidding when he said, "It takes a tough man to make a tender chicken."

Avian flu has played a major role in toughening up the poultry business too. When turkey farms unfortunately located under the migratory pathways of wild birds in California, Minnesota, and Ontario came down with bird flu in the 1960s, the industry responded by moving the birds indoors. When the disease hit chicken jails in the 1980s, many factories adopted an "all in, all out" system to keep the virus out. Instead of trucking the fowl around to different feeding locales, they now raised meat birds in one spot until slaughter. "They keep changing farming practices to keep ahead of avian flu," explains Brown.

Before the Asian outbreak, avian flu had "officially" visited the poultry industry only 21 times since 1959. The most memorable invasion took place in 1983, when a mild strain of H5N2 invaded

Pennsylvania's chicken factories. After it mutated and started to swell chicken brains, authorities ordered the destruction of 17 million birds at a cost of $61 million. The epidemic affected three states and raised the price of chicken across the country. South of the border, it also raised hell. In an attempt to stay ahead of the invader, the Mexican authorities set up a rigorous poultry vaccination program. In spite of the program (or, some argue, because of it—many of these vaccines were substandard), a mild strain of H5N2 slowly morphed into a highly pathogenic killer that plucked the feathers from Mexico's chickens. The progenitor of this invasion jumped into the wild birds of Mexico, where it smoulders, waiting for another dinner. Meanwhile Mexico's industrial chickens have required 2 billion doses of vaccines in the last two decades in order to make it to the supermarket.

Avian flu isn't the only frequent microbial visitor at chicken factories. In recent years every Tom, Dick, and Harry of the avian disease world has come calling on industrial piles of meat. Six-hundred-page books and entire journals with dry titles like *Avian Diseases* are now devoted to a seemingly endless list of chicken predators, including cholera, fowl pox, infectious coryza (a sinus clogger), Marek's virus (a tumor maker), infectious bronchitis virus (a SARS relative), and retroviruses that behave like AIDS. Given that the movement of people, equipment, and vehicles can effortlessly spread these invaders, it's not surprising that many factory-minted pathogens are beginning to spill over into wild birds too.

In 1994 American bird lovers discovered wild house finches with crusty eyelids and impaired vision at backyard feeders. A common factory poultry scourge, *Mycoplasma gallisepticum* (MG), had attacked them. Scientists have yet to identify its source. The epidemic wiped out tens of millions of songbirds and spread like wildfire throughout the eastern seaboard. (Blinded by the infection, the hapless finches couldn't see to eat.) The plague's remarkable speed so impressed scientists that the journal *Emerging Infectious Diseases* declared that MG "illustrates how rapidly a pathogen can be disseminated throughout a large geographic area within a highly gregarious and mobile host population." Then came the Asian outbreak of avian flu.

It all started with mountains of poultry. In the last decade Asia's domestic bird population jumped from 4 billion to 16 billion. Between the 1970s and the mid-1990s, the volume of animal protein devoured in China and environs grew three times faster than it did in North America and Europe. As the middle classes of Hong Kong, Shanghai, and Mumbai got more cash in their pockets, they wanted more chicken in their bowls. Sitting down with a pile of meat, after all, is a badge of progress toward good living.

But Asia's burgeoning chicken trade ignored some basic epidemiological facts. For starters, Asia has been the epicenter of influenza pandemics for thousands of years. You can't crowd people, pigs, and poultry together without building a better market for viral commerce. China's Guangdong province, which virologists call the "birthplace of influenza," teems with 86 million people, tens of millions of pigs, and about 250,000 birds per square kilometer (about 650,000 per square mile). One public health expert recently told the *Los Angeles Times,* "Charles Darwin could not have set up a better genetic re-assortment laboratory if he tried." Since 1900, Guangdong has probably been the seedbed of two human influenza pandemics. But it's not the only flu nursery. When pigs, wild fowl, and crowded U.S. military camps turned Iowa and Kansas into mini-Guangdongs in 1918, the flu that sprang from the American midwest infected half of the world's population and killed more than 50 million people.

Robert Webster, a prominent influenza scholar at St. Jude's Children's Research Hospital in Memphis, Tennessee, has repeatedly warned the world's chicken growers that they shouldn't fool with the flu. In particular, the virologist has explained that feeding pigs and chickens in the same neighborhood assists viral exchanges; that live bird markets, where as many as 276 bird species mingle, give flu viruses ample opportunities to re-sort themselves; and that commercial poultry farms offer "an optimum number of susceptible poultry for rapid viral evolution."

Michael Osterholm, a well-known American public health advocate, has made a similar point. In a 2005 issue of the *New England Journal of Medicine,* Osterholm noted, "It is sobering to realize that in 1968, when the most recent influenza pandemic occurred, the virus emerged in a China that had a human population of 790 million, a pig population of 5.2 million, and a poultry population of 12.3 million. Today these populations number 1.3 billion, 508 million, and 13 billion respectively."

But the chicken trade has made more friends in government than it has among grumpy doctors obsessed with overcrowding. Take Thailand, for example. It banked on the broiler chicken to elevate its economy and make it "kitchen to the world." Before its avian ambitions transformed the tropical nation into a global chicken exporter, most peasants depended on hardy native fowl for pocket change and protein. These farmers kept, on average, about 15 hardy chickens, geese, and quails that foraged freely for food. The owners didn't fuss about medicated feed, vaccination, or medication. The birds also mixed with their wild migratory cousins, ensuring that avian influenza repeatedly immunized their stock with mild strains of the disease or occasionally culled them with deadlier versions.

However, Asia's livestock revolution changed that bucolic scene by escalating viral evolution several notches. In Thailand the government and the Charoen Pokphand Group (CP Group), a conglomerate run by ethnic-Chinese entrepreneurs, adopted the bigger-is-better model, or what the Asian media call the "Guangdong syndrome." Inspired by American-style poultry factories, the CP Group promoted "chicken farming estates" in closed buildings that housed between 10,000 and 100,000 birds each. The CP Group lent money to farmers, sold them chicks, feed, and antibiotics, and bought the finished birds to export to Japan and Europe.

Outside of Bangkok, a megalopolis of 10 million chicken eaters, the volume of domestic fowl slaughtered for export grew from 410 million to 890 million birds in just one decade. Most of this expansion took place in poultry factories. Thailand, a relatively small country, miraculously became the world's fourth-largest poultry exporter. Similar chicken concentrations have piled up outside Hanoi, Manila, Guangzhou, Shanghai, and every major Indian city.

Not all the Asian economies chose chicken as the miracle maker. China and Vietnam invested in ducks and boosted production threefold in the last two decades. The two countries now house three-quarters of the world's ducks. According to Samuel Jutzi, director of the FAO's Animal Production and Health Division, this swarm of 1 billion ducks and geese unfortunately developed on less "than .5% of the earth's terrestrial surface," where excess phosphates from chicken waste had already polluted 30 percent of the farmland. He's not at all surprised that the cackling concentration of domestic geese and ducks provided "an effective breeding ground for the myriad of avian influenza viruses circulating in wild waterfowl."

Although government and industry officials still pin the origin of the Asian pandemic on wild fowl, most scientific experts at the FAO don't think it spilled from "the flyways to the highways and byways." Migratory birds have surely played a role in transporting the virus (as have the poultry trade and live markets), but the pandemic didn't begin with them. A weighty 2005 report for the FAO by two New Zealand epidemiologists concluded that the virus probably "arose through a recombination process between viruses in the influenza epicenter region of Asia involving interchange of viruses among a number of species of domestic birds." In other words, all the trouble started on the byways and highways around factory birds and then flooded wild flyways.

The big viral party probably began in the 1990s when highly pathogenic strains of avian influenza struck commercial poultry factories throughout southern China. In 1996 domestic geese started dying in heaps in busy Guangdong province, where millions of people and billions of fowl live cheek by jowl. It then spread to domestic ducks and eventually spilled over to broiler chickens and quails bound for Hong Kong, one of the world's most densely populated cities. When the indomitable Hong Kong virologist Yi Guan later traced H5N1's lineage back to China, his lab was temporarily closed for violating "state secrets."

The first human casualties appeared in 1997 as avian influenza ripped through the Hong Kong's poultry operations, infecting 18 people. After six citizens dropped dead, the city declared an emergency and mobilized 2,000 government workers to stamp out the plague. In what would soon become a repetitive scene, men in ghostly moon suits bagged and gassed the birds at 160 farms and 1,000 wet markets and buried the fowl in landfills. Hong Kong then banned all aquatic fowl from its markets and mandated two cleanup days a week, when the markets would be scrubbed down and disinfected. China maintained its sorry record of secrecy and issued the usual polite denials.

H5N1 created quite a buzz among virus researchers. It reproduced much faster than its innumerable relatives, killed wild birds (something avian influenza shouldn't do), and mowed down domestic chickens more quickly than any avian invader encountered before. It had also jumped species and appeared to have pandemic potential. H5N1 wasn't supposed to infect humans, let alone kill them. "We thought we knew the rules," Stephen Morse, director of the Center for Public Health Preparedness at Columbia University, later told *Newsweek*. "And one of those rules was that H1, H2 and H3 cause flu in humans, not H5. This is like the clock striking 13."

But H5N1 wasn't the only avian invader making headlines. In March 2003 the Netherlands, a country with its own "Guangdong syndrome," got a visit from one of H5N1's less virulent cousins, H7N7. The Dutch epidemic began after free-range chickens picked up the virus, probably while mingling with some wild swans on a pond in the center of the country. The virus then hitched a ride on contaminated litter, vehicles, and the clothing of poultry workers. As it worked its way through uniform chicken "units," the virus heated up and became a chicken eater. More than 400 farm workers and slaughtermen exposed to infected poultry got a bit of the flu and came down with conjunctivitis and the sniffles. One veterinarian died when pneumonia filled his lungs with fluid. The Dutch called out the military and slaughtered 30 million chickens on 255 premises, nearly one-third of the nation's 100 million birds. Police even raided the homes of pet lovers, who tried to hide their beloved chickens in the bathroom. To protect the chicken killers, the government handed out Tamiflu, face masks, and goggles. The barriers did not prevent an outbreak of depression, anxiety, and "emotional aggravation" among farmers and vets. The virus spread to Germany and Belgium too, where another 3 million birds got the ax.

Back in Asia, the feared H5N1 strain continued to make the rounds in China's congested poultry factories. It took up residence in healthy domestic and wild ducks in the coastal provinces and southern cities. It, too, did some gene swapping with pigs. When scientists inoculated domestic chickens and mice in 2003 with samples of the latest version of the virus, their laboratory patients dropped dead. So too did ferrets, a species with a high tolerance to flu. "Clearly," concluded the understated researchers, "H5N1 influenza viruses are continuing to evolve in Asia."

The Chinese inadvertently helped the virus along. First, they started to douse chicken factories with the human antiviral drug amantadine. This explains why H5N1 mutated and has appeared in drug-resistant forms throughout Asia. Then they hit their city-sized flocks with a sloppy and frenzied vaccination program that used bad batches of inactivated virus from 20 different medical facilities. "Initially they were not matching the virus to the vaccine," explains the virologist Earl Brown. These poorly matched vaccines simply drove H5N1 underground. Although the birds looked healthy, they still shed the virus in feces, infecting unvaccinated birds in "silent epidemics."

The Mexicans conducted a similar uncontrolled experiment in viral evolution in their flocks from 1980 to 2004. Their industrious 20-year vaccine campaign tempered the virus but did not eliminate the disease. In fact, the *Journal of Virology* recently reported that the strain merrily kept on evolving under vaccine pressure—a development that scientists call a source of "concern for the effectiveness of vaccination strategies within the poultry industry." (Incredibly, the use of Mexican-made vaccines in Japan has now been linked to several outbreaks of avian flu.)

Thanks to China's haphazard interventions, H5N1 learned some new genetic tricks and started to kill other fowl in 2002. It flattened a flamingo, an egret, two gray herons, and other waders in a Hong Kong zoo. Within short order it was killing ducks, something an avian virus had never done before. Pigeons and house sparrows soon began to pile up in Hong Kong and Bangkok.

In August 2003, evolution hit a new high in Thailand, Indonesia, and Vietnam as a highly pathogenic strain (genotype Z H5N1) started to explode through factory farms and backyard operations. (Indonesia, like most governments, didn't officially recognize the epidemic for another six months.) This latest mutation of H5N1 directly attacked chicken brains. Some birds laid misshapen eggs before dying; others flopped around like drunks. Most just dropped dead.

In Vietnam the first wave unsettled large commercial farms along roads. A second wave in March 2004 devastated the Mekong Delta, a poultry and human beehive. In Thailand the virus jumped from free-grazing ducks on rice ponds into chickens and back again. In Indonesia it raced from island to island. According to the Indonesian Poultry Information Center, an illegally imported vaccine from China may have helped to spread more silent epidemics of H5N1.

Throughout Asia despairing farmers protected their avian assets as best they could. In Indonesia, where the famed Kampung chicken is every family's cash cow, ordinary people concealed their troubles, one poultry official explained later. "Farmers feel shame if their birds are infected by AI [avian influenza], [so] they try to keep the information to themselves. Farmers also feel scared that if government knows they have AI, they must stamp out their birds and have no pay. So they try to sell the birds as soon as they can in order to save their capital." Not surprisingly, wet markets became distribution points for the bird flu.

By now the virus was taking advantage of every man-made opportunity and poultry habit. Major festivals, such as the Chinese New Year and Tet, brought birds and people together in markets in huge numbers. In Thailand fighting cocks, players in the country's most popular sport, transported H5N1 from village to village. A good rooster can command as much as $20,000, so millions of Thais predictably hid their fighting birds from murderous government officials intent on protecting the export trade.

In the midst of the avian pandemonium, H5N1 startled virologists once again by jumping species. After it killed 45 tigers and clouded leopards in Thailand, zookeepers belatedly removed raw chicken from their diets. When infected tigers passed on the disease to other tigers, authorities euthanized another 102 animals. Captive tigers, leopards, and lions died in heaps in Cambodia and China. By now H5N1 had shown "the widest host range" of any avian virus ever seen. In Holland researchers proved that household cats fed raw chicken could transmit the virus too. During a human pandemic, cats could become silent spreaders.

Then people started to drop dead, just like ducks and tigers. In a small village in Thailand where all the local chickens succumbed, 11-year-old Sakuntala Thongchan developed a fever and stomachache. She coughed up blood until viral pneumonia suffocated her lungs. Her 26-year-old mother died of avian flu shortly afterward. In Vietnam three brothers toasted one another with glasses of duck blood; one died, one fell ill, and another lived to tell the tale. After slaughtering a chicken, a 21-year-old farmhand named Nguyen Si Tuan battled the flu for 82 days and lost a third of his body weight. In Thailand a cockfighter sucked the mucus from a bird's throat in order to revive his prized possession and died a week later. In Indonesia entire families perished after slaughtering old laying chickens. And in southern Vietnam a four-year-old boy and his sister died of diarrhea and acute brain swelling after bathing in a duck pond. British researchers marveled at H5N1's versatility: "The virus is progressively adapting to mammals and becoming more neurologically virulent."

In a belated effort to control the pandemic (which now affected 10 Asian countries), governments orchestrated the killing of millions of birds in early 2004. Bare-handed workers, farmers, and soldiers all became reluctant chicken killers. In Vietnam they burned chickens alive; in Thailand they packed 11 million birds into fertilizer sacks and buried them alive. To get rid of the "evil spirits" affecting their fowl, Indonesians barbecued flocks on pyres. The poorest of the poor, however, did what they've always done: They ate their diseased livestock.

To help out, the United Nations printed pamphlets on chicken disposal that offered this useful nugget: "It can be extremely difficult to remove dead birds from their cages once rigor mortis is established." Infectious disease experts watched the slaughter on their television screens with alarm: "They are trying to eliminate the animal reservoir, which is what we want, but if they are exposing themselves to the virus while they're doing that it might defeat the purpose," declared one official at the World Health Organization.

Many of the world's influenza experts were almost beside themselves. An influenza virus must meet three conditions to start a wave of human trouble. It must be novel—something to which people don't have immunity. It must kill with ease, and it must spread easily from person to person. H5N1 had now fulfilled the first two conditions. The flu expert Robert Webster repeatedly called H5N1 the "worst flu virus I have ever seen or worked with." Greg Poland, an infectious-disease expert at the Mayo Clinic, likened the whole affair to "watching a train wreck in slow motion." Most of Asia's birds, wild or domestic, had become viral factories that manufactured mild or deadly strains of H5N1 for up to 17 days. The virus was everywhere.

In addition to taking advantage of a chronic shortage of vets, laboratories, public health experts, and doctors, H5N1 thoroughly exploited Asia's imbalanced chicken economy. The region's 200 million backyard operations, which are annually seeded with flu strains by wild fowl influenza, sit next to high-tech commercial factories where feed shipments, workers, egg trays, and live birds move in and out every day. This commercial traffic juggled the virus from backyard operations to factories and vice versa. "The commercial broiler chains may well have played a major role in the seeding" of the virus, according to FAO livestock ecologists. To date the owners of Asia's factory farms have refused to open their disease records for public scrutiny.

As poultry prices plummeted and peasants started eating rats in Vietnam, relatives of H5N1 prompted more killing around the globe. In August 2004 South Africa started culling $30 million worth of infected ostriches while Maryland officials dispatched 4 million birds. South Korea and Japan also battled outbreaks. After Taiwan dispatched hundreds of thousands of chickens and ducks, an unusually frank editorial in the *Taiwan News* lamented the wave of plagues pummeling the region, from SARS to foot-and-mouth disease: "Regrettably, these major epidemics all came from China. Why? The most important reason is trade and smuggling of agricultural and livestock products."

Europe learned that lesson in October 2004 when a Thai bird smuggler went on a business trip to Antwerp. In a cloth sports bag he carried two rare eagles bought at a bird market in Thailand for a European collector. Had it not been for a routine drug search, the birds would have escaped detection. Both were infected with H5N1. The incident opened a small window on a large and illegal global enterprise: the smuggling of 4 million pet or exotic birds around the world every year. Shortly afterward Taiwan found eight infected birds in a load of 1,000 smuggled birds from Fuzhou, China. And in Port Elizabeth, New Jersey,

U.S. Agriculture Department agents seized 12,000 kilograms (27,000 pounds) of smuggled frozen poultry from China.

By now most of the world had witnessed the remarkable geopolitical prowess of avian influenza. When H7N3, a mild relative of H5N1, invaded poultry factories in the Fraser Valley of British Columbia in the spring of 2004, the Canadian government rolled up its sleeves, called out the killers, and culled 19 million birds. The Canadian Food Inspection Agency, a booster of livestock trade, tried gassing the flocks with carbon dioxide. When many birds recovered and walked again, they clubbed them with sticks. They shot peacocks with shotguns and poisoned some duck populations two or three times before they died. The government stamped out rare breeds as well as millions of uninfected broilers. In the interests of international trade, Canada's bird killers elected to kill first and test later.

During the slaughter, neither government nor industry talked about the perils of overcrowding or high-density poultry production. But British Columbia's chief medical officer of health, Perry Kendall, pointed out the obvious: "The intensely farmed birds tend to be very genetically similar. The methods of farming result in them being actually more frail and more vulnerable to diseases, particularly since there are so many of them in such a small volume of space."

Back in Asia, H5N1 continued to gobble up birds and people while governments did what Mark Dekich did: They protected trade and tried to avoid the word *influenza.* The French reporter Isabelle Delforge aptly described Thailand's handling of bird flu as "a saga of cover-ups, incompetence, lies and extremely questionable decisions." The nation's top microbiologist, Prasert Thongcharoen, later told *The New Yorker* that the government's response had been an unmitigated fiasco. "They didn't do the right thing. I'm not saying it would have stopped an epidemic, but they didn't do what they should have done."

For starters, the country's poultry kings and the Department of Livestock (two largely indistinguishable entities) denied the existence

THE CHICKEN EXPLOSION

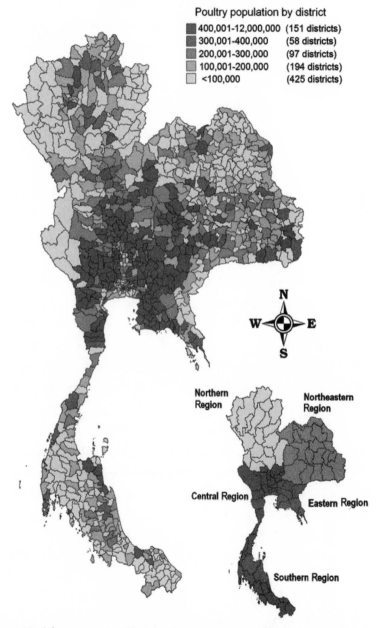

Poultry population by district
- 400,001-12,000,000 (151 districts)
- 300,001-400,000 (58 districts)
- 200,001-300,000 (97 districts)
- 100,001-200,000 (194 districts)
- <100,000 (425 districts)

Northern Region
Northeastern Region
Central Region
Eastern Region
Southern Region

Viral factories: Many Thai districts support more poultry than people.
Source: Centers for Disease Control. http://www.cdc.gov.

of the disease even after a veterinarian found H5N1 in November 2003. Trade-conscious officials explained at first that the sick chickens were dropping dead "without any medical cause." Meanwhile infected birds were traveling to market by the thousands. Then the government, fearing public panic and trade sanctions, claimed it was dealing with a ho-hum case of bird cholera, a disease treatable by antibiotics. But still the chickens died. The birds would start shivering en masse, then they'd fall over dead. The lies and delays, however, gave industry ample time to capitalize on the rising prices for frozen chicken parts.

When the truth finally came out and the country's $1-billion chicken trade collapsed, Thailand's president, Thaksin Shinawatra, did what politicians do best: He publicly ate well-cooked chicken. "If Thais don't eat Thai chicken, how can we expect others to buy our chicken?" cooed another panjandrum. The government sponsored chicken festivals and the CP Group gave away thousands of free chicken dinners. While pop stars sang sweet songs about "the kitchen of the world," public relations experts encouraged Thais to eat at Kentucky Fried Chicken outlets, arguing that the fast-food chain served only factory chickens and that factory chickens were virus-free. As European trade bans left piles of raw chicken meat on Thai docks, the government tried to broker an unusual deal with Sweden: poultry for a JAS 39 Gripen jet fighter. Sweden politely declined.

Officials looking for scapegoats found a convenient one in wild birds, the natural reservoir for the virus. Thai leaders gave orders to butcher wild Asian open-billed storks, and the trees they nested in were hacked down. In Singapore the order went out to cull crows and mynahs and to trap and clip the wings of migratory birds. People shunned nature reserves and impulsively released their pet songbirds. In Hong Kong authorities cut down more trees so birds could not nest in schoolgrounds. In Taiwan educators forbade children to visit nature reserves, and in Japan they shot crows. Colin Poole, Asia director for the Wildlife Conservation Society, had never seen anything like it. "Birds had become the enemy."

Although migratory ducks, geese, and swans arguably provided H5N1 a free ticket out of Asia, wild birds didn't spread the invader around China, Vietnam, or Indonesia. A 2006 genetic analysis of migratory birds and factory chickens in southeast China confirmed that the invader started in Guangdong, colonized local areas, and then spread via the local poultry trade. Wild birds are "not the scapegoats for maintaining H5N1 within poultry," the Hong Kong researcher Malik Peiris told the *New Scientist.* "There the cause and solution lies within the poultry industry."

For the last two years, most wild fowl carrying H5N1 have been found dead or dying near chicken factories, and, as many observers have noted, "the dead don't migrate." (Ducks appear to be an exception; they can carry the virus with little misery.) Many bird species targeted by irrational governments, such as open-billed storks, never carried the virus in the first place. As Poole adds, "no waterfowl spend the winter in southern China and then migrate further south." In sum, "the international movement of poultry and other birds in trade" initially spread a virus lethally served up by factory farms. Yi Guan, the astute Hong Kong virologist, has called the migratory bird issue a "free lunch" for governments that can't or won't control legal or illegal movement of poultry or pet birds. "Each time there's an outbreak they say, 'It's migratory birds. I cannot control them. I cannot lock my sky!'"

To many scientists the relentless slaughter of chickens no longer makes any sense on ethical, economic, or ecological grounds. The virus is now here and, given the structure of the modern poultry business, is not about to depart anytime soon. Given the paucity of good viral monitoring and detection, no one really knows how many of the 200 million birds killed, burned, or gassed to date were uninfected or would have recovered. "These animals, whether chickens, ducks or pigs, have lives too, and almost no religion on earth would condone such massacres of living organisms on such a scale, especially with preventative measures," wrote three indignant Chinese scientists in *Nature.*

Dave Halvorson, a University of Minnesota veterinarian and flu expert, also doubts the modern poultry business can "tolerate this expensive, unproven, draconian and dramatic method of disease control much longer." The need for a continuous supply of meat and eggs, he says, has caused many poultry factories "to act in ways that do not contribute to disease control and may actually contribute to disease spread." Although Halvorson argues that it is no longer necessary to "consider diseased or convalescent poultry as evil," global traders and governments have not accepted the message. Nobody, of course, wants to talk about crowding.

Earl Brown, who has spent much of his career studying how population densities govern the virulence of viruses, doesn't understand why chicken numbers still haven't become a priority. "Neither governments nor the media want to talk or hear about it." Without proper research on population dynamics, chicken factories will simply hatch more viral trouble. "We have to know what's happening and figure out the calculus that drives virulence, so we don't repeat it."

Yet the global menu calls for more chicken in greater and greater concentrations. The Great Chicken Pandemic has even strengthened its most efficient incubator: the factory farm. The pandemic has displaced thousands of farmers and is actively replacing the backyard chicken with so-called biosecure corporate operations. Singapore has banned all small-scale poultry farming, as have several cities in Vietnam. In fact most jurisdictions with factory birds want to ban free-range chicken farming or small backyard poultry flocks as threats to "biosecurity." Even Elizabeth Krushinskie, president of the American Association of Avian Pathologists, now regards organic and free-range chicken as a risk to "the increased prevalence of highly concentrated food animal agriculture." Many Asians now suspect that the lively traditional banter of the region's colorful wet markets will eventually be supplanted by corporate slaughterhouses and the Muzak of superstores.

Meanwhile H5N1 continues its quest for global citizenship. In the spring of 2005, the Year of the Rooster, H5N1 killed off 6,000 bar-headed geese on Qinghai Lake in northern China, nearly 10 percent of that species's population. It then started to gobble through domestic fowl across the country, hitching a ride on untreated poultry manure, smuggled birds, and poultry trucks. After hundreds of outbreaks and probably scores of human deaths, the government announced an amazing plan: the vaccination of 14 billion birds. No one knows what quality of vaccines will be used or whether the vaccination campaign will drive the virus to greater lethal mutations. Yi Guan, the combative Hong Kong truth seeker and virologist, doesn't even think the scheme is feasible. "This is a very crazy, a very stupid idea." He thinks the only way to end the invasion is for all infected countries to shut down their chicken factories and start over.

When migratory fowl started to die in wetlands and lakes in Siberia, Kazakhstan, and Mongolia in the summer of 2005, Europeans panicked. Sweden clamped down on Tamiflu hoarders, Holland and France moved their commercial birds indoors, and the Croatian government ordered its citizens not to eat raw eggs. In England trigger-happy hunters picked off migratory birds flying over the Channel. But in London bird importers blithely mixed consignments of quarantined birds, transferring H5N1 from a batch of Taiwanese exotics to parrots from Suriname and proving once again that migratory birds aren't the only viral carriers.

South Koreans experimented with sauerkraut and kimchee, a fiery cabbage dish, to treat fluish birds. In Japan authorities traced 31 outbreaks of H5N2, a mild relative of H5N1, at chicken factories to an illegally smuggled vaccine from Mexico. As H5N1 insinuated itself into every nook and cranny of the poultry trade, many scientists abandoned their Chicken Little predictions and got philosophical: "We're just in chapter five of a 20-chapter book on avian flu—there is so much we don't understand," reflected one exasperated Indonesian veterinarian.

H5N1 isn't the only member of the influenza tribe looking for diversification opportunities. Pigs, for example, used to harbor two stable influenza subtypes (H1 and H3), but in the last decade, "pig genetics has gone spinning in wild directions," says Brown. Now avian, human, and porcine types are all swapping genes in the world's expanding pork factories. To date the viral invasion hasn't caused much more than a lot of porcine sneezing. "But we have a really varied viral soup in pigs right now," notes Brown.

Dogs haven't been so lucky. With absolutely no warning, a horse flu (H3N8) jumped to greyhound racing dogs in 2004. (Greyhound pups are traditionally fed horse meat.) The outbreak quickly burned through a dozen U.S. states, including New York and Florida. The new canine influenza virus starts with an unusual "kennel cough," then proceeds to a bloody pneumonia for about 10 percent of puppies and older dogs. Canines and felines can also carry H5N1.

None of this frantic viral activity in mammals surprises the famed Dutch virologist Jaap Goudsmit. He argues that viruses are now breaking out of their natural reservoirs in wildlife at unprecedented rates, thanks to persistent human encroachment. Humans now utilize 83 percent of the earth's surface, commandeer 60 percent of the freshwater, and consume a third of all living things in the ocean. That doesn't leave much room for other creatures. "A new phase seems to have begun in the evolution of avian flu viruses," Goudsmit muses in his book *Viral Fitness*. Given that human activity or what we call globalization is rapidly eliminating birds and mammals at nearly a thousand times the natural rate of extinction, Goudsmit wonders, "Where else other than among humans can a virus family settle down in the coming centuries?" Where else, indeed?

Most experts expect that H5N1 will soon invade North America, where the World Bank thinks it could cause a multibillion-dollar headache for the fast-food business. They also predict that avian flu will deliver an economic calamity in Africa, where backyard chickens are the only capital many women can own. "It's out of the box now," admits Henry Niman, an American flu expert and president of Recombinomics, a biotech firm. Every day Niman publishes an insightful commentary on the elegant viral evolution of H5N1 on the Internet (http://www.recombinomics.com/). Most end this way: "The dramatic expansion of H5N1 in poultry and people is cause for concern as are reports of dead dogs and cats." Niman believes that H5N1 is now so widespread that it can't help but become a human virus. "The only question is, How virulent will it be?" He doesn't think antiviral drugs or vaccines will slow a pandemic down much. He marvels at the weak public health response, the shoddy monitoring, the repeated deception by governments, and the limited number of effective interventions. "People are talking about the avian flu but they're not doing much about it."

The talk has been colorfully pessimistic. Mike Leavitt, U.S. secretary of health and human services, describes the world as a "biologically dangerous place right now." Lee Jong-wook, director general of the World Health Organization, says, "The next pandemic will cause incalculable human misery." The United Nations secretary general, Kofi Annan, warns that "we would have only a matter of weeks to lock down the spread before it spins out of control." Jaap Goudsmit worries that influenza might jump to cattle, while the French microbiologist Antoine Danchin fears we are just reaping what we so dutifully sowed: "It is impossible to separate infectious diseases from our lifestyle or from the structure of our societies, and above all, from venal considerations. Our infections mirror our primary interests and our way of life."

In all of its manifestations, avian flu has reminded men in white coats that influenza is what it is. It moves more like a hurricane than like an invading army. It can act faster than governments or the peripatetic professionals of the World Health Organization. It can hijack a chicken factory, travel with manure into ponds, and take over a flock of migratory birds. It can spread unseen, thanks to the oldest of human passions:

trading and smuggling. And it can out-evolve vaccine producers or drug makers. It is, in short, a biological storm gathering velocity and force.

The feathered blizzard of influenza also possesses many ancient advantages. It can travel along the most intimate of pathways: a handshake, a kiss, a cough, or a tear. It can swim in water and sleep in frozen duck meat. It can silently infect poultry workers without illness, leaving only a ghostly antibody behind as a calling card. And it can noiselessly sweep through any crowd. Once a few particles enter the human lungs, influenza multiplies furiously. Within a couple of days, the sick are shedding tens of millions of particles in nasal mucus alone. When an airplane full of globetrotters lost its ventilation for about four hours in 1977, influenza ran through the passengers at lightning speed: nearly three-quarters of them got the flu from just one sneezing passenger.

At a 2005 symposium organized by Deutsche Bank, the well-known influenza researcher Robert Webster wrung his hands as he has done many times in the past. What began as sneezes and low-grade fevers in Guangdong's poultry factories in the 1990s has become, just a decade later, a distinct serial killer of ducks, chickens, and migratory birds. When given to ferrets, animals with a legendary tolerance for influenza, it caused diarrhea, forced breathing, leg paralysis, and death. "This is the first influenza virus that kills ferrets and spreads throughout the body to the brain." Although the virus has killed many human chicken handlers and colonized thousands more with silent infections, it has yet to acquire a real knack for person-to-person transmission with pandemic ease. Webster has no doubt that the evolving virus will master that trick next. "I think that we have to accept the fact that we are sleeping on top of a time bomb."

So far that's the good news. The really bad news is that there is more than one improvised explosive device ticking along the global roadside.

TWO

Juggling Species:
The Global Circus

*"Humanity is the purposeful but often drunken
ringmaster of a three-ring circus of organisms."*
—Alfred Crosby, Jr.

A vian flu's remarkable blitzkrieg of the world's bird populations
wouldn't have fazed Charles Elton. The shy Oxford ecologist often
biked to class in the 1950s and typically arrived bedecked with leaves
from a morning trip to Wytham Wood. For most of his life he avoided
human crowds, preferring the company of animals. But as the father of
animal ecology, an expert on Arctic lemmings, and a patient observer of
2,500 different creatures in Wytham near his home, Elton knew a thing
or two about biological storms. He called them simply "explosions."

In 1957 the middle-aged professor boldly declared during a three-
part BBC Radio series called *Balance and Barrier* that "we live in a very
explosive world." Thanks to bigger trade and faster transport, ever-
rational *Homo economicus* had tipped the natural balance of things by
transgressing every geographic barrier and then some. By uprooting
species of plants, animals, and microbes that evolved in distinct realms
and setting them loose willy-nilly, humans were fooling with evolution

in a dramatic way. In fact, invaders, the biological equivalent of break-and-enter gangs, now threatened to rearrange every household of nature with one frightful explosion after another. "I use the word explosion deliberately," wrote Elton, "because it means the bursting out from control of forces that were previously held in restraint by other forces."

The prescient scientist recognized that invaders set off two very different kinds of explosions. In the first bang, an alien species arrives in new terrain with the assistance of human hands. Free of old parasites and diseases, the intruder takes over with Mafia-like determination. The chestnut blight, for example, didn't have much of a career in Asia, where local trees had developed good defenses against it. But when the nursery trade introduced the voracious fungus to New York forests in 1902, it took off like the avian flu. In just 50 years, the invader spread over 91 million hectares (225 million acres) and killed a quarter of the shade trees in eastern U.S. forests, nearly 4 billion trees. Ten moth species that lived on the chestnut tree became extinct, and a good part of North America died forever.

But not all invaders are imported aliens. Sometimes climate change or human meddling can set off native explosions, Elton's second cate-gory, by quietly altering local food supplies. The Colorado potato beetle held a discreet position in the American southwest as a minor plant muncher until a human invasion plowed up the local greenery and replaced it with a monoculture of Peruvian interlopers: the potato. The beetle couldn't believe its good fortune and tucked in like a Las Vegas gambler at a free buffet. It now dines on most of the world's global potato plots.

At the time of the radio series, no one took Elton's invasion tales to heart. A year later, the ecologist published a short, witty book on the subject: *The Ecology of Invasions by Plants and Animals*. Full of elaborate military-style maps that tracked the triumphant campaigns of European starlings or the Mongolian bubonic plague, it warned that biological invasions could create many "lost worlds." Although Elton's conclusions alarmed a small herd of biologists, the book didn't become a best seller. Shortly afterward, Rachel Carson's exposé on chemical pollution, *Silent Spring,* grabbed the public's imagination and Elton's book was all but forgotten. But while environmentalists started to rail against pesticides and PCBs, "great historical convulsions in the world's flora and fauna" silently unsettled more landscapes.

Over time it became apparent to many biologists that Elton had identified the bigger and uglier threat. By 2001 even the General Accounting Office, a congressional watchdog, was admitting that invasive species are "one of the most serious, yet least-appreciated, environmental threats of the 21st century." Chemical spills, although messy and harmful to food chains, have a beginning and an end. Biological invaders aren't so neat; they have a nasty habit of reproducing; they usually expand their territorial claims over time; and they often adapt to local conditions by making hybrids. Biological invasions, in other words, tend to last forever. Daniel Simberloff, a longtime invasion chronicler at the University of Tennessee, calls invaders "strangers in paradise" or "infectious diseases of the environment."

Our long and cultivated ignorance of biological invaders helps to explain why ordinary folk are now surfing the Internet looking for flu remedies. It might also explain why an army of diverse invaders embargoed a third of the world's meat trade in 2003 and why nervous doctors swarm to congresses dedicated to "emerging diseases" that appear at a stunning rate of one every eight months. When Elton penned his warning in 1958, the United States was dealing with approximately 70 global swarmers; today more than 7,000 microengineers are remaking American forests, crops, and waterways. They are not only reordering trade but also subverting politics, rattling health officials, kicking empires, and digging graves. For better or worse, invaders are rearranging our kitchens, bathrooms, and bedrooms—forever.

The medical establishment, a trade of slow learners, still hasn't grasped the whole crazy picture about biological manipulation on a global scale. Most physicians still think that invasive bacteria, viruses, and fungi threaten the health and sex lives of only the world's urbanites. In fact, most white coats don't understand that the same global forces driving "emerging diseases" are also delivering entire armies of "emerging" interlopers to our livestock, crops, forests, wildlife, and water. But professionals tend to be turf protectors and reductionist thinkers by training. The big picture often eludes them and us. But if zoologists, veterinarians, plant pathologists, oceanographers, and epidemiologists actually talked to one another, they could all swap Elton-like invasion tales.

Animal doctors could regale the crowd with outrageous stories about mad cows, rinderpest, foot-and-mouth disease, and avian flu or the whole sorry unprecedented and unparalleled explosion of livestock diseases. Next, plant pathologists would lament the invasion of citrus trees by canker bacteria, the fungal assault on bananas, or the latest invader of vineyards. Oceanographers would then add to the chaotic picture by documenting the rising number of die-offs consuming every form of coastal marine life from turtles to lobsters. Foresters would chip in with campfire tales about massive forest threats, including invasive fungi, wildfire beetle epidemics, and exotic tree species. Wildlife biologists could tell a yarn or two about chronic wasting disease in elk and deer or how a curious fungus has undone global frog populations. Finally, public health experts might wrap up with sober PowerPoints on AIDS, anthrax, multiresistant germs, and other human invaders.

Such frank exchanges could end with only one obvious conclusion: that all invaders live by the same damnable code. Whether dining on trees, animals, or New Yorkers, invaders all hitchhike along familiar global pathways: ships, planes, and goods. Once tourists secure a beachhead, they proliferate like crazy, and wait for a chance to explode. Although many invasions fail, successful invaders belong, like humans, to the global fraternity of the fast and the furious. It's pretty hard to beat organisms as fast as avian flu or as fertile as purple loosestrife, a single plant of which can produce 2.7 million seeds a year.

Like international merchants, invaders are opportunists; both groups have a singular disdain for barriers and borders. With the energy of Wal-Mart executives, they often construct a new big-box world more vulnerable to disaster and extreme events than the original ecosystem. Invaders are the world's original deconstructionists, and their eventual product is a sort of McDonaldization of creatures. "Invaders are tragically engineering a total homogenization of the earth's flora and fauna," notes David Schindler, a former student of Elton's and one of the world's foremost water ecologists. "Elton wouldn't be surprised, but he would be depressed by it all."

Perhaps the most melancholy part of this great reshuffling is what scientists call the sixth great extinction. They predict that our grandchildren will see 50 percent fewer mammals, birds, and butterflies than we did. Michael McKinney, a geologist at the University of Tennessee, says it's all about replacing many losers with a few winners. The losers tend to be sensitive locals with specific geographic allegiances that don't tolerate city making, climate change, or forest cutting very well. They include elephants, tigers, parrots, fish, and other species that breed slowly. In other words, the losers are often rare, big, and not overly fond of humans. The invasive winners, however, have all the traits every successful globalist needs: They are small, fertile, and omnivorous, and they adapt to any messy landscape or cityscape shaped by human hands. They include ticks, avian influenza, rats, methicillin-resistant *Staphylococcus aureus,* cryptosporidium, cholera, African cane toads (the bane of Australia's outback), white pine rust, and just about every fungus you can name. If unchecked, says McKinney, this "biotic homogenization" will result in fewer and simpler landscapes and will probably kill off more species than the Great Rubout at the end of the dinosaur age.

Scientists, the masters of neutral language, are loath to place all invading creatures in one black book (doctors study microbial invaders while biologists study animal and plant invaders), but "an invader is an invader is an invader," says Hugh MacIsaac, a prominent Canadian biologist. Avian flu and Asian tiger mosquitoes invariably depend on human mobility and exploit mundane pathways of trade. The traffic in animals, logs, food, and ornamental plants, accidentally or intentionally, furthers the adventures of invaders. Some hitchhike in wooden packing crates while others stow away in used tires. Others lie hidden in boxes of vegetables, unprocessed logs, and nursery plants. Many microimmigrants arrive as minnows in a bucket, fungi on a plant's roots, seeds in a pocket, frozen meat in a suitcase, or germs in human feces. Others come with price tags and can be bought in pet stores. Economists, notes MacIsaac, mostly forget or ignore that global trade is just what it is: "It is trade in everything."

The Great Lakes, which make up the world's largest body of freshwater, darkly reflect how a global tradefest can turn into an ecological makeover. What was once a distinct North American body of water full of Arctic grayling and Atlantic salmon is now little more than a degraded multicultural aquarium. In the last 200 years, 182 alien fish, mollusks, algae, fleas, and plankton have invaded and re-sorted the chemistry and biology of the Great Lakes. Every six months a new immigrant jumps in and starts a new aquatic subdivision. The globalists hail from just about every corner of the planet: New Zealand, Russia, Africa, the Atlantic, China, and Great Britain. Not surprisingly, all have arrived with the same evolutionary goal of making a better life for themselves.

The Great Lakes bestiary now includes a Caspian fishhook water flea, the Atlantic sea lamprey, the tube-nose goby, and the New Zealand mud snail as well as the infamous zebra mussel. Some of these organisms sailed with cargo ships while others arrived in the discarded contents of aquariums. Many swam up man-made canals and locks that connect the five lakes to Atlantic trade routes. Faced with few enemies and blessed with extreme mating habits (and the new global order definitely favors the fertile), these invaders have done what opportunists, colonists, and entrepreneurs do everywhere: They've redecorated the place to their own liking without malice aforethought.

Many new arrivals have behaved with all the civility of pillaging Vikings or rampaging Mongols. Great Lakes invaders have fouled water pipes; slimed fishing nets; murdered scores of native fish, invertebrates, and plants; encrusted boats; stunk up beaches; and even changed the basic biology and chemistry of the lakes themselves. (The Great Lakes, you might recall from geography class, are glacier-carved features visible from the moon. So changing the chemical makeup of 243,000 square kilometers [94,000 square miles] of water is no mean feat.) Edward Mills, a biologist at Cornell University and a longtime lake watcher, just scratches his head. "It's becoming more difficult to understand who's eating what and who's impacting what."

Most of the 35 million citizens living on the Great Lakes aren't aware of what's happening in their aquatic backyards. As relative newcomers themselves, they don't have any long memories of the lakes or how they evolved or used to be. (Think of shores of white pine, waters full of Atlantic salmon and whitefish, and skies full of passenger pigeons—all gone.) The idea that the Great Lakes might be a blue metaphor for global dislocations changing water bodies, islands, and continents worldwide doesn't yet resonate with people. But if they knew that a quiet 9/11 was happening in their drinking water, they would probably be stunned if not alarmed.

A majority of the interlopers have arrived in the last 40 years, and trade explains why. Since the opening of the St. Lawrence Seaway in 1959, the Great Lakes region has become a global hot spot and the world's sixth-largest economy. Ports such as Detroit, Thunder Bay, and Toronto annually handle more than 250 million tonnes (275 million tons) of cargo, everything from cars to wheat. Vessels from 460 ports on five continents do business on the lakes every year. The majority of trade still comes from northern Europe, which is home to 70 percent of recent Great Lake invaders.

The zebra mussel, a striped, fingernail-sized mollusk from brackish waters flowing into the Black Sea, is the lakes' most celebrated invader. (It is to North American waterways what avian flu is to chickens.) Unlike most hitchhikers, it has actually won a little media attention, for the simple reason that it has become a multibillion-dollar headache. The story of how a Russian native is now doing to the Great Lakes what the plague did to 14th-century Europe—that is, changing everything— is a classic invasion tale.

The invasion began sometime in the mid-1980s and was entirely unannounced. At the time, wheat crops had failed in the Soviet Union, forcing the struggling regime to send empty ships to load up on Canadian prairie wheat stored in terminals along the Great Lakes. These grain carriers took on ballast water in European ports such as Rotterdam, already colonized by the hardy mussel. To steady the ship on high seas, the grain vessels pumped aboard enough water to fill several hundred Olympic-size pools. But collecting ballast water is like netting catfish: It's a messy job. In addition to brackish water, the ship sucks up sediment, worms, larvae, cysts, viruses, and other guck. Members of a local mussel gang hitched a ride. Once the ships reached grain ports on the lakes, they flushed out the ballast water as well as the other biological stowaways, which in recent years have even included cholera. History is made and unmade by such trivial events.

Every year global shipping moves 80 percent of what humans eat and buy and in the process discharges 3 billion to 5 billion tonnes (3.3 billion to 5.5 billion tons) of ballast water all around the planet. (About 50 million tonnes [55 million tons] of these trade excreta ends up in the Great Lakes.) As a consequence, scientists estimate, more than 7,000 different species of marine microbes, jellyfish, plants, fish, and fleas are relocating every day. Ships' ballast has become third-class accommodation for enterprising aquatic invaders. Although many stowaways don't survive the first or even the second trip (most invasions fail), repeated invasions can and do create a series of unforeseen and unfortunate events.

The zebra mussel, *Dreissena polymorpha,* was one of them. Once flushed into the lakes, the Black Sea invader found what French and English immigrants discovered in the 1700s: a paradise with few predators, unlimited food, and lots of lovely real estate. The mussels had established colonies in Lake St. Clair, between Lakes Huron and Erie, by 1988. Since then they have hitchhiked on boats, anchors, and other paraphernalia to invade the Mississippi River, the Missouri River, and the inland waters of 20 American states.

THE BALLAST WATER CYCLE

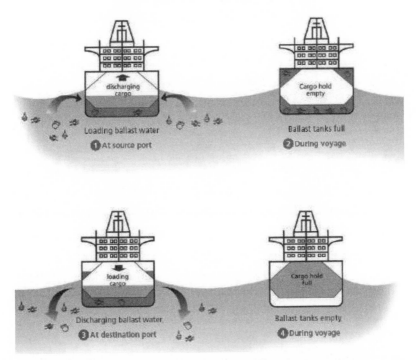

Ballast water transports more than 7,000 invasive species around the globe every day.

Source: GloBallast Programme, IMO. Reprinted by permission.

The invader's miraculous breeding habits abetted this Napoleonic expansion. So too did its ability to cling to any hard surface, including the bottoms of thousands of recreational boats. A single adult female can squirt out 1 million eggs a year, and each offspring can reach sexual maturity in 12 months. In just one square meter (11 square feet) of shoreline, one happy colony increased its numbers from 1,000 to 700,000 individuals in six months flat. Like suburban housing developers, the mussel fully understands the importance of exponential growth.

The zebra mussel has now infested the Great Lakes so completely that scientists agree the invasion is irreversible. The Black Sea colonist not only covers most hard surfaces along the shorelines but also has encrusted native clam beds with smothering colonies of thousands of mussels. These swarmings generally rob the locals of vital nutrients and, together with pollution, have exterminated as many as 32 species of indigenous clams in one part of the Great Lakes alone. Whenever the mussels run out of clams, anchors, boards, shipwrecks, and rocks to overwhelm, they just encrust themselves in layers 20 centimeters (8 inches) thick. It's not unusual for swimmers and beachcombers to cut their feet on the razor-sharp shells of mussels and have to be hospitalized for tetanus.

But that's just the beginning of the mussel's imperial rumblings. As industrious filter feeders, the mussels vacuum up all the local algae, changing the clarity of the lakes. Every day one mussel can filter one liter or quart of water, and there are now enough mussels in Lake Erie alone to filter the entire volume of the lake once a week. Thanks to the diligent screening of billions of mussels, Erie is now four to six times clearer than it was before their arrival. Light-sensitive species, such as walleye, are now looking for sunglasses. In addition, nearly 140 native clams are facing extinction. "Zebra mussels have changed the course of these lakes forever," notes Mills. "They will never march back."

The mussels have not attracted much attention as exterminators of native wildlife. They won some notoriety only when they started clogging intake and outflow pipes for power stations, factories, and water utilities. The mussels like to grow inside pipes because these environments provide good shelter and a constant supply of food. Zebra mussels can fill two-thirds of a water intake, sluice, or irrigation canal in short order. Some towns on the lakes have even had to impose water rationing because of mussel infestations on intake pipes. Nearly a dozen nuclear power plants now spend millions of dollars every year cleaning up the growths on their critical cooling systems. Pipe cleaning is a costly business. The U.S. Fish and Wildlife Service estimates that the zebra mussel invasion will cost taxpayers more than $5 billion in basic infrastructure damage over the next 10 years.

But the zebra mussel isn't the only ballast invader reshaping the Great Lakes. Its cousin the quagga mussel, another Ukrainian import from the Black Sea, is also fouling up water pipes and killing off native clams. The spiny water flea, a native of Great Britain and northern Europe, arrived in the late 1980s and is now competing with native fish that eat zooplankton. Like the zebra mussel, the spiny water flea is a fecund invader; its females can even produce young without mating. Its cousin the fishhook water flea arrived in 1998 from the Black Sea and is already notorious for robbing food from the cradles of larval fish and frosting fishing nets with its ghostly prickly skeleton.

These new arrivals help one another, just as human immigrants do. The clumped shells of the zebra mussel provide shelter for a Black Sea crustacean, *Echinogammarus ischnus*. It also feeds on the excrement of zebra mussels. Wherever mussels rule the lakes, the biomass of *E. ischnus* has increased 20-fold. So too have outbreaks of botulism, which has killed ten of thousands of loons and ducks.

These aquatic epidemics illustrate the prowess and unpredictability of invader synergies. The bacteria that produce botulin toxin accumulate first in the guts of birds that have eaten round gobies. The gobies primarily feed on zebra mussels, which do a fine job of concentrating the bacterial toxins (and chemical ones) while filtering lake water. Billions of mussels produce copious excrement, which in turn creates the perfect conditions for the explosive growth of *Clostridium botulinum,* a bacterial manufacturer of botulism par excellence.

"People think we've seen it all. But we haven't seen the half of it yet…. The chance of a serious problem increases exponentially with each species that we add," says Anthony Riccardi, a researcher at McGill University. He calculates that another 40 species are poised to invade the lakes, including a so-called "killer shrimp" from Europe. "It's one of the few things in nature that will kill much more than it needs to eat."

The synergy of invaders is nothing new. Before the zebra mussel rewrote the chemistry of the Great Lakes, the sea lamprey, an ugly eel-like creature, rewrote their biology. It did so by attaching itself to native fish and sucking the life out of them. Introduced to the lakes in the 19th century by a series of canals to the Atlantic Ocean, the parasite didn't really take off until the 1930s, when fishermen started to pull up lake trout and whitefish with two or three gorging lampreys attached. Between 1945 and 1950, the volume of lake trout catches on Lake Michigan alone fell from 2.3 million kilograms (5 million pounds) to just 2,300 kilograms (5,000 pounds). The lamprey wiped out three fish species: the Atlantic salmon, the lake trout, and the longjaw cisco.

The disappearance of these predators allowed the alewife, a silvery 15-centimeter (6-inch) fish and another Atlantic interloper, to reproduce without fear. Schools of alewives proved more adept at eating native plankton than did either native herring or whitefish. As a consequence alewives now make up 90 percent of the biomass in Lake Michigan. Fish managers quickly saw the alewife as a convenient food source for the Pacific salmon, an alien introduced to replace the native predators wiped out by the sea lamprey. The salmon dined on the alewives, but they passed on an enzyme that drains salmon of thiamine; without this vitamin, the sport fish are doomed to shorter life. And so the lakes that the human settlers Jacques, Jack, Mario, Ali, Ching, and Nasar built keep on getting stranger.

Some biologists now describe the biological assault on the Great Lakes as an "invasional meltdown." Over time the sheer numbers of the invaders and their unpredictable interactions could exterminate all the original denizens of the Great Lakes. Once invaders have hit an ecosystem hard, it becomes more vulnerable to successive invasions. In fact, invaded landscapes are to global biological opportunists what failing states are to terrorists: boomtowns.

So living on the lakes is no longer a North American experience. It's now possible to catch fish from Europe and Asia (carp and brown trout), step on mussels from the Black Sea, get stuck in massive beds of milfoil from Asia, collect the odd piranha from the Amazon (releasing aquarium pets is a common occurrence on the lakes), be bitten by mosquitoes carrying West Nile virus from Africa, and watch maple trees die from the destructive work of the Asian long-horned beetle. Why tour the globe when trade, immigration, and travel will bring the world (and its hitchhiking barbarians) to your doorstep?

The relocation of species has been part and parcel of evolution for a long time. But everyone agrees that *Homo economicus,* the global trader, has greatly speeded up things. Before standardized 18-tonne cargo containers (usually called 20-ton containers) ruled the world, a wind-storm or a migrating bird might have dropped a new insect or microbe on the Great Lakes every 50,000 years or so. But the human quest for more gadgets, vacations, and food has dissolved all temporal and spatial boundaries. In many respects, human history reads more and more like a string of biological explosions that grow ever more loud and grotesque as the pace and scale of our trading increases. Elton called it "an unending string of unforeseen emergencies."

Globalization probably began about 10,000 years ago with the invention of agriculture in the Middle East. After humans domesticated cattle and wheat, we couldn't wait to spread the good news around. Seeds, plants, and animals—and their bacterial and viral hitchhikers—have been on the march ever since. The first exchanges took place between Africa and Eurasia and probably included a few viral and bacterial explosions that emptied many a promising new settlement. The Babylonians, the Egyptians, and the Greeks, all passionate collectors and traders, dealt in plants and animals with increasing regularity.

By the time the Romans turned the Mediterranean into *mare nostrum,* Marcus Aurelius was trading African rhinoceros horn for Chinese cherries. In fact, the Romans imported so much alien flora and fauna that more than 400 African and Asian perennials can still be found around the grounds of the Colosseum, where traffic in foreign gladiators and wild animals reached a bloody intensity. Meanwhile traders of the Han Dynasty in China sailed with sorghum to East Africa and brought back camels; then they opened the Silk Road, a formidable pathway for invaders. In Africa the Bantu rearranged the southern half of the continent with millet, cattle, and goats. In fact, every rising culture and empire has traded away its own favored creatures, its "portmanteau biota." Aborigines arrived in Australia with the dingo, while Polynesians refused to leave home without a good assortment of pigs, yaro, and yams.

The constant traffic in wildlife, plants, and foods picked up speed during the Tang Dynasty in the 7th and 8th centuries. The Tang, one of China's richest and most poetic empires, traded as far afield as Asia, Europe, and Africa. During the 13th and 14th centuries, the Mongols took up the challenge of globalization with a single-mindedness later envied by World War II generals. Their well-traveled desert highways eventually helped transport the plague from Asian marmots to Europe, where the grisly invader exploded and eventually buried nearly a quarter of that continent's population. Reviewing such well-recorded explosions, the American virologist Stephen Morse has concluded, "Globalization can be a very good thing, but there are careless by-products."

The intensity of careless by-products hit a new peak in the 1400s with the so-called Columbian Exchange. The erudite American historian Alfred Crosby, Jr., has argued that Christopher Columbus and his heirs started the equivalent of a biological war when they stumbled upon the New World. Blessed with a distinct biology, the Americas looked like a bubble boy to Europe's seasoned and disease-ridden biota. After Columbus unpacked his bag, one explosion after another transformed the landscape. The Norwegian rat, the Hessian fly, the French dandelion, and what we now call Kentucky bluegrass ran riot. A variety of microbial opportunists or microinvaders also tagged along on imported livestock.

The immune systems of aboriginal people in both North and South America had never seen anything like it. Smallpox, measles, and tuberculosis swept through the locals with the energy of zebra mussels and buried much of the New World's genetic diversity. When disease failed to kill the people off, feral European cattle, pigs, or sheep ate them out of house and home. In Mexico alone, the population crashed from a pre-Conquest high of 25 million to 1 million souls in just two centuries. All in all, the Americas lost 90 percent of their local inhabitants. "Bringing previously isolated ecosystems together is much like flicking a cigarette lighter near open containers of gasoline," notes Crosby. "Some of the time nothing will happen. Some of the time fumes will ignite and blow your head off."

Meanwhile European plants, crops, and livestock filled in the abandoned real estate colonizing American grasslands. Invader just helped invader. Cattle grazed down native plants to their roots and broke the soil. Thistles, nightshade, sedge, white clover, and cheatgrass soon followed. Most of the New World, including the Americas, South Africa, and Australia, received a thorough biological makeover. British flora and fauna (nearly 55 exotic mammals were introduced by Europeans) replaced native New Zealand wildlife so completely that locals still call the country "Britain's offshore farm."

The decimation of aboriginal people by European invasive diseases started a chain reaction. Faced with chronic labor shortages for planned sugar plantations and silver mines, the Europeans turned to Africa and revved up the slave trade. That forced-relocation scheme prompted another set of disturbing and remarkable exchanges. With the importation of millions of African slaves came malaria, yellow fever, West African rice, and the African honeybee. The latter import interbred with native bees in Brazil to become the notorious "killer bee."

All of this unprecedented traffic in germs, crops, livestock, and people created what we now call the global marketplace, says Crosby. "For the first time human labor, raw materials, manufacturing and transportation systems spanning scores of degrees of longitude and latitude by sail, wheel and beasts of burden were organized on a world scale." Thanks to the new globalism, one in four citizens in Amsterdam in the 16th century hailed from tropical climes. Since then globalization has shrunk the world from baggy bloomers to a thong, and the pace of trade and traffic has just gotten faster, with only a few sobering trade slowdowns, such as the Great Depression in the 1930s.

Along the way invaders got help from different quarters. In the 19th century, European "acclimatization societies" started to move around lots of foreign flora and fauna. These associations of inventive globalists didn't like the strange look of their new colonies or of their own backyards, for that matter. They wanted, as one French society proclaimed, what most globalists now take for granted: "to endow our entire society with blessings which have been neglected or unknown until now." Many impossible dreamers openly aspired to put an Asian yak in every French vineyard, a Japanese chestnut in every American forest, a European rabbit in every Australian pot, and a Brazilian water hyacinth on every pond. Driven by dreams of commerce, vanity, and just plain silliness, the acclimatizers planted starlings in New York, European brown trout in California streams, nutrias (South American rat-like rodents) in Louisiana, and weasels in New Zealand. An attempt to herd kangaroos into the American West fortunately failed. Aided by seed merchants and nursery trade, the acclimatizers spread prolific weeds, such as Japanese honeysuckle, tamarisk, and kudzu, into the U.S. mainland and Hawaii and beyond.

Today the world doesn't need acclimatizing societies anymore. The speed, democracy, and efficiency of our global economy can turn any individual with a bucket or an aquarium into an ecological imperialist or biological bomb thrower. A fascinating 2005 study by the British fish expert Gordon Copp neatly illustrated the human penchant for species juggling. Copp began his study by draining a number of ponds in the Epping Forest north of London. (Fish, by the way, are England's third most popular pet, after cats and dogs.) He diligently removed all the goldfish and other global swimmers, then restocked the watering holes with native fish. Next he watched and waited to see if new invaders would arrive by chance or by "environmental or human factors."

Copp didn't wait long. Within a year all the ponds within just 10 meters (11 yards) of a road or footpath filled up with an average of 3.5 varieties of koi carp, orfe, shubunkin, goldfish, or other Asian ornamental imports. People dumped the creatures because they got too big, too diseased, or too boring. In one four-week period alone, people abandoned 15 alien fish in Jubilee Pond. Copp's conclusion: "Chance as a natural phenomenon appears to play little or no role." Humans, in other words, make not only careless species collectors but also remarkable species dumpers.

What's happening in England's ponds is happening pretty much everywhere we trade, play, or work. In the 1980s, global trade outpaced all previous trade since the beginning of civilization. Planes, ships, and automobiles now accidentally or intentionally do to hundreds of species every hour what acclimatizers took years or months to do. The value of food trade alone, for example, has grown by 200 percent in the last 20 years, from $100 billion to $300 billion, and much microbial traffic has tagged along. Not surprisingly, the incidence of food-borne invaders and food-borne disease has increased around the world by 50 to 100 percent. Three-quarters of the world's citizens now consume vegetables and fruits from foreign sources. (This explains why food-borne illnesses now sicken 40 million North Americans a year and kill another 10,000 people.) Microbial hitchhikers are nothing if not persistent: New Englanders now drink coconut milk and cholera from Asia, Norwegians nibble lettuce and shigella from Italy, and New Yorkers dine on canned mushrooms and staph bacteria from China. A group of food safety specialists working for the World Health Organization wrote: "Over the past 200 years the average distance traveled and the speed of travel have increased 1,000 times while incubation periods for diseases have not changed."

Modern shipping, of course, has outdone Columbus and Marco Polo by opening really broad pathways for invaders. In the 18th century only a handful of sailing ships could transport more than 270 tonnes (300 tons) of cargo. By the 1960s, the average vessel had scaled up to 5,400 tonnes (6,000 tons). Today most container ships haul approximately 5,000 20-ton containers full of shoes, shirts, clock radios, and bananas. (The shipping industry boasts that it has enough containers to build a 2.5-meter-high [8-foot-high] wall around the equator twice.)

Invaders can't resist a free ride. Between 1986 and 1990, Chinese inspectors intercepted 200 species of weed seeds in 349 ships in Shanghai alone. When an Australian entomologist recently studied the detritus in 3,000 unloaded cargo containers, he found more than 7,000 insect stowaways. Some of the interlopers were dead and some were alive. They included formidable tree killers, crop eaters, fire ants, and other global troublemakers, proving once again that trade is trade in all things. During 11,000 inspections in 2003, New Zealand customs agents found more than 600 different organisms, including fungi, geckos, spiders, and "miscellaneous animal material." A third of the species were novel immigrants.

Invaders hitch rides not only in cargo containers but also in wooden packing material, including pallets, blocks, skids, and dunnage. Probably more than half of all trade goods that arrive in North America are protected by wooden shipping crates. That amounts to about 15 million individual crates every year.

The green, often bark-covered wood of these crates hides a plethora of aliens. About 80 percent of the non-indigenous beetles and pathogens trying to invade the continent's already pest-ridden and blighted forests arrive in pallets. Between 1985 and 2000, three-quarters of the beetles outed by U.S. plant inspectors had used wooden crates for transport.

The volume of biological traffic on wooden crates has grown exponentially. A 1997 Canadian audit of 50 wooden spools bound with Chinese wire located a secret cargo of seven different forest chewers in just a quarter of the spools. In 1998 a shipment of Norwegian granite wrapped in green spruce arrived in Montreal and then traveled by rail to Vancouver. Horrified Canadian forest inspectors later discovered 2,500 insects representing more than 40 different species of foreign tree eaters and diseases in the dunnage. At least three were major forest invaders with the same expansive appetite as Dutch elm disease. It, too, arrived on wooden crates in the 1930s.

Ballast water has also been a formidable medium for strangers seeking to invade local paradises. As early as 1981, a typical ship entering the Great Lakes without cargo nonetheless brought along as many as 150 species of phytoplankton and 56 different invertebrates in its ballast. That scale of invasive introductions simply put 19th-century ships—with their dry ballast of earth and rock—to shame. Elton recounts how a tropical researcher in 1930 counted a mere 41 dry species on a Rangoon rice ship, including cockroaches, fleas, and rice beetles.

A single 747 jet can transport dozens of the 1,415 known human pathogens as well as ticks, fleas, mosquitoes, and parasites to any global destination in just hours. (Catching "airport malaria" from vagabond mosquito invaders isn't all that unusual at busy international airports.) A single car can harbor 3,926 seeds representing 124 different plant species. Trade and transport now have the power to collapse political borders and erase geographical ones.

In the global economy, both domestic and wild animals are constantly on the move too. Every year about 350 million creatures get bounced about in the international pet trade. Between 1989 and 1997, the United States imported more than 18.3 million iguanas, snakes, rodents, geckos, hedgehogs, tropical fish, birds, and lizards. Every year hapless Americans searching for novel home entertainment import nearly 2 million reptiles from 80 different countries. Indonesia exports 25 tonnes (23 tons) of turtles every week. Some reptiles become pets, while others are transformed into medicines and shoe leather.

The traffic in wildlife may someday make the Columbian Exchange look like a Mickey Mouse affair. Since 1992 the number of wildlife species—everything from pythons to pangolins—sold legally in pet stores has jumped from 200,000 to 352,000. This beastly trade has dismayed Robert A. Cook, chief veterinarian for the Wildlife Conservation Society, and for good reason: "The result is a dangerous concatenation of circumstances with animals and would-be consumers from different ecosystems coming into contact," he warned the U.S. Congress in 2003. "The staggering numbers of animals and people coming into contact with each other change the one-in-a-million odds of disease spillover into almost a daily possibility."

Most consumers don't realize it, but every time they frivolously buy an exotic pet, they are really buying more than one alien. Macaques, for example, truck along herpes B; Amazon parrots hump a good pack of Exotic Newcastle disease; African leopard tortoises tote ticks bursting with heartwater fever, a plague of livestock; Gambian giant rats carry a payload of monkey pox (yes, a relative of smallpox); African clawed frogs harbor a deadly fungal pathogen; and 90 percent of all lizards, snakes, and turtles transport salmonella. Not surprisingly, this blood poisoner and bowel emptier now invades 74,000 American reptile lovers every year.

The $25-billion illegal trade in exotic pets, which threatens a third of the world's birds and mammals with extinction, is equally friendly to invaders. It too serves as a conveyor belt for SARS, Q fever, and other marvels unknown to science because most wildlife pathogens haven't been identified. "It is apparent, therefore," recently wrote the American veterinarian Corrie Brown, "that the traffic in non-traditional species is robust, all creating the meteorological forces of a perfect microbial storm.... Animals and people are inextricably interconnected like the lines in an Escher drawing."

Human travel has also increased exponentially right along with all that reptile mobility. In 1965 only 75 million of the world's 3 billion people considered themselves global citizens. Today more than 230 million live far from home, flitting about like starlings. In fact, 1 in every 35 persons now lives abroad in some teeming port city. (If gathered together, they would make the world's fifth-largest country.) Tourists are multiplying like zebra mussels. Every year 400 million people visit Europe, and by 2010 nearly 1 billion tourists will don Tilley hats and digital cameras to board jet planes in the increasingly endangered hope of finding something distinct or native in our globalized planet.

So the world is looking more and more like a global zoo and behaving like one, too. The planet now boasts some 1,200 menageries and they, of course, trade specimens. Most of the 2 million creatures caged in zoos in 72 countries are bred in other zoos. To Brown this translates into a "plethora of possibilities" for hitchhiking invaders: "An African rodent born and raised in an Asian zoological collection can be found in a South American zoo housed next to Arctic mammals from North America." But now we all live in zoos.

Brown compares the rapid emergence of invaders driven by global trade to an old carnival game called "Whack the Mole." Every time a toy mole pops its head out of one of a myriad of holes, the player whacks the rodent with a rubber mallet—and another immediately surfaces. "The satisfaction derived from neutralizing one mole is immediately replaced with the drive to beat back the next mole that appears out of one of many holes." But Brown doesn't think *H. economicus* has any better strategy for dealing with invaders than a reactive one: "beating back each new disease as it puts its head out of the hole."

As one of the world's predominant exporters and importers, China has become another professional mole whacker. Biologists have learned that the most reliable way to chart and predict the pathway of invaders is by checking trade routes. Since 1980 trade between China and the United States has risen from $5 billion to $231 billion. That makes China America's third-largest trading partner. The partners, however, are trading more than TVs, plastic goods, toys, and clothes.

In 1999 American traders sent red turpentine beetles on a shipment of raw logs to China. The California native has since eaten its way through 10 million pines by fouling up their circulatory system and starving them of water and nutrients. At about the same time, Chinese traders sent a collection of emerald ash borers to Detroit on a wooden pallet. A bunch of Asian long-horned beetles arrived in Toronto in wood packing materials after having colonized Chicago and New York. The two Asian invaders have since chewed their way through about $6 million worth of trees around the Great Lakes and could eventually cause $138 billion in damage to the timber industry as well as leave many cities denuded.

China now calculates that explosive growth in trade and transportation is setting off $7-billion biological explosions every year in its livestock, crops, and forests. "These problems are bound to become even more expensive as China increases its trade ties with the United States and other countries," one forest mandarin recently confided to *The Columbus Dispatch*. But most Americans still don't understand why an endless parade of undeclared Chinese immigrants continue to hog the headlines: the Asian tiger mosquito, SARS, the northern snakehead fish, and, of course, avian flu. After the "Beijing Exchange" will come the "Mumbai Exchange," because every trading empire has practiced its own ecological imperialism.

One of the things that really bothered Charles Elton was the disproportionate attention given to invaders that killed, as opposed to those that just sabotaged economies. The killers, whether they filled morgues or barns, always dominated the news. While plague and influenza became well-known invasive celebrities, the chestnut blight and the Asian long-horned beetle labored in obscurity. Even though they made it "more expensive to live," the economic terrorists got little respect.

They still do. Global trade, in fact, seems spectacularly blind to its biological costs. The potato fungus, now on a second global tour, has cost billions and threatens food security for the poor. Two beef invaders, foot-and-mouth and mad cow disease, have racked up more than $30 billion in bills for European and North American livestock industries. The boll weevil, a Mexican free trader, arrived in the 1890s and has since run up a $50-billion tab for the U.S. cotton industry. Dutch elm disease arrived on a shipment of European logs 70 years ago and has since killed 46 million trees, or $20 billion worth of timber. Invasive weeds cost Australian agriculture $3 billion a year. In Texas, South American fire ants eat their way through $300 million worth of livestock and public health budgets annually. All in all biological invaders now cost the North American economy approximately $137 billion a year.

Few if any of these outrageous economic robberies ever show up on any government or corporate ledger. Nor has any auditor general brought global traders to task for ruining local economies with deliberate or accidental invasions of influenza, ants, or water hyacinth. The reasons for such deadly oversight are simple, says Daniel Simberloff. For starters, importers and exporters don't hold biology degrees. But the real problem always comes down to accountability. The costs of invasions are borne not by the merchant class but by consumers. The global health bills for avian flu are sent not to poultry factories but to taxpayers. In a recent essay on the politics of risk assessment for invaders, Simberloff explains, "A substantial benefit that might accrue to a few has more political weight than a substantial cost that might be borne forever by all."

Conflict of interest also explains why global traders ignore the health costs of biological invasions. The folks in charge of monitoring the introduction and spread of noxious invaders are usually the same folks charged with facilitating exports, imports, and interstate commerce. "This is a profoundly schizophrenic mission," notes Simberloff. It is also a profoundly global predicament. The idea that we should close the door on invaders challenges a global fad: "the touting of free trade as the solution to social and economic ills."

With trade on their side, most invaders stay in the shadows, much like illegal aliens in any global city. The Argentine ant stowed away on coffee shipments and unpacked in southern California decades ago. It's now displacing native species as well as starving out horned lizards, which can't eat it. But Governor Arnold Schwarzenegger has no plans to terminate them. A Japanese vine, kudzu, has smothered so much land in the U.S. southeast (2.8 million hectares, or 7 million acres) that even poets have accepted the invader as "a mindless unkillable ghost." In Puerto Rico the Indian mongoose gobbles up local frogs and snakes the same way big-box stores eat up small businesses. But no anti-mongoose lobby speaks of unfair biological trade. Thanks to the intentional release of smallmouth bass and red slider turtles by aquarium owners, Japanese rivers wouldn't seem all that unfamiliar to Mark Twain. But in the burbling Tamagotchi culture of modern Japan, no one expresses much regret about the loss.

Elton offered an eloquent recipe to avoid catastrophic explosions, but no one has really adopted it because it would mean slowing down and shopping less. He proposed that people who wanted to drain a wetland, plow up native grasses, chop down a forest, or traffic in exotic goods should ask themselves three questions: How will this change affect the animals and plants that live here now? What beauty and security may be lost? And how will this change affect "the accumulating instability of communities"? With a few exceptions (Australia and New Zealand), no country has bothered to ask those questions. Nor do traders or politicians. The American social critic Wendell Berry argues, "Arrogant ignorance promotes a global economy while ignoring the global exchange of pests and diseases that must inevitably accompany it."

Whether arrogant or ignorant, species jugglers tend not to ask a lot of questions. And so we adapt to BSE in our hamburgers, emerald ash borers in our trees, cholera in our water, zebra mussels in our intake pipes, *C. difficile* in our hospitals, cane toads in our fields, purple loosestrife in our wetlands, and avian flu in our nightmares. Chris Bright, a researcher with the Worldwatch Institute, recently lamented, "Humanity is not meant to be a patient in a sickroom with nothing to contemplate but our own diseases." But that's what invaders are making us.

Joshua Lederberg, a Nobel prize–winning biologist and long-term germ watcher, says that humans are now doing to the world something they've never done before. "We have crowded together a hotbed of opportunity for infectious agents to spread over a significant part of the population. Affluent and mobile people are ready and willing and able to carry afflictions all over the world within 24 hours notice. This condensation, stratification and mobility is unique, defining us as a very different species from what we were 100 years ago."

This means that the ultimate invader, *Homo economicus,* must now grapple with a homegrown war of the worlds. Aliens have definitely broken into our house, but we opened the door and provided the transport.

Each and every day our trading habits now set off the biological equivalent of improvised explosive devices in every global port and megacity. After the smoke clears, the only important question left is this: Will *H. economicus* be one of the few winners replacing the many losers?

THREE

Livestock Plagues

*"A monster is no more than a combination of parts of real beings,
and the possibilities of permutation border on the infinite."*
—Jorge Luis Borges

P eter Chalk, a policy analyst with the Rand Institute, told the U.S. Congress in 2003 that bioterrorism could easily toast the country's $50-billion agricultural export industry. Then the diligent consultant rhymed off the vulnerabilities one by one. For starters, most modern livestock have little disease resistance, thanks to "steroid programs and husbandry practices instituted to elevate the volume and quality of meat production." Packed in feedlots or gestation crates, most livestock in the United States are so stressed by overcrowding that their natural resistance to viral and bacterial infections has been shattered, explained Chalk. What's worse, 22 highly contagious exotic diseases, such as African swine fever and avian influenza, had recently become global menaces. All these nasty invaders are easy to either acquire or spread, added the biowarfare expert. The sheer concentration of animals (the average dairy facility now boasts as many as 5,000 to 10,000 animals) guarantees that an infectious outbreak would be difficult to contain and "could necessitate the wholesale destruction of all the animals." In sum,

Chalk noted that a terrorist could easily set off a livestock plague that could cost billions, unsettle governments, and even "create mass panic, particularly if the catastrophe had a direct public health impact."

In his insightful presentation, Chalk neglected to mention that global trade had already delivered such biological disasters in spades and now threatened the kin of Bessie, Porky, and Chicken Little with more diseases than history has ever seen. Before a different audience, Bobby Acor, the former chief of the U.S. Animal and Plant Health Inspection Service, echoed essentially the same warning: "We're only one sandwich, one suitcase, one cargo container away from disaster."

The scale and scope of wild plagues eating up livestock might grant new meaning to the age-old practice of giving prayerful thanks for dinner. In the last decade alone, the Netherlands has called out the troops to slaughter 30 million chickens, and Taiwan's gross national product dropped two full points after a devastating pig plague. Mad cow, a multibillion-dollar trade disaster, has beaten up the cattle industries of Europe, Canada, the United States, and Japan. Foot-and-mouth disease, a rather mild virus that affects neither meat nor man, gutted England's tourism industry and resulted in the irrational slaughter of 10 million animals. Nipah virus, a pathogen carried by fruit bats, killed 100 Malaysians in 1999; it cost the pig industry $350 million and resulted in the slaughter of 1 million domestic porkers. Postweaning multisystem wasting syndrome (PMWS), a viral invader of porcine immune systems, also has the world's pork industry in a state of alarm. Since 2003 Asians dressed in moon suits have gassed more than 200 million chickens, ducks, and quails in an attempt to control an avian influenza pandemic that could give the world a fatal case of flu. It seems that only a combination of dumb luck and brute force continues to keep meat on the global table.

Many meat invaders are turning up in herds for the first time. In northern Africa, Libyans have spent $80 million battling an infestation of New World screwworm, a creepy flesh-eating fly that normally terrorizes livestock in South America. Rift Valley fever, a deadly African resident carried by mosquitoes that causes bleeding in humans, is now poised to invade not only the livestock of the Middle East but its keepers as well. Other countries are busy reading up on a well-named cast of troublemakers, including bluetongue virus, sheep pox, bovine TB, and Crimean-Congo hemorrhagic fever. The world's livestock herds—and our dinner—are under full-scale biological assault.

To date these unprecedented invasions have cost the global economy more than $100 billion and resulted in the slaughter of more than 1 billion animals. With the growing densities of livestock populations around the world and the fact that the majority of human infectious diseases come from animals, more plagues are coming. Our penchant for cheap meat clearly comes with hidden and huge biological costs.

Vaclav Kouba has been documenting the fallout. He's a respected Czech animal scientist and the former head of the UN's Animal Health Service at the Food and Agriculture Organization (FAO). Dismayed by the carnage, he now calls the huge stream of diseases decimating livestock what it really is: "a crisis of veterinary medicine." How can animal doctors say they are upholding their historical mission to promote and protect animal health when global trade now threatens to create "a man-made irreparable disaster" in animal and human health, he asks.

The major driver for these deadly exchanges has been the phenomenal growth of the global meat trade. Every day the volume of live animals as well as steak, pork chops, and chicken wings crossing borders breaks another record. In the last 40 years, the amount of meat moving about has increased from 3.2 million tonnes (3.5 million tons) a year to 24 million tonnes (26 million tons), calculates Kouba. At the same time, the number of animals in transit has grown to 40 million a year. In fact, livestock and meat now account for nearly a tenth of all global trade. While Europeans dine on Thai chicken, Saudi Arabians enjoy Australian beef, and Greek diners chow down on British lamb. Japan's rice eaters chomp poultry from Brazil, and in the United States diners gobble up shrimp from China. But this globalization of meat has brought other guests to the table.

According to Kouba, 607 imported diseases have hit the world's livestock herds since 1980. More cases of avian influenza and foot-and-mouth disease have popped up in the last decade than in the previous century. About a third (212 cases) of these invaders have the potential to cause deadly diseases in people, such as avian flu and Rift Valley fever. More than half of the invaders appeared in countries they had never visited before. Most of the invaders proved impossible or difficult to control, and almost all arrived legally, with international paperwork certifying them as disease-free. Global disease reporting is incredibly lax (no country likes to report animal disease, let alone human epidemics), so these numbers reflect the tip of the iceberg.

Kouba, who has written exhaustively on the spread of animal diseases, blames this mayhem squarely on the quest to maximize trade profit at the expense of animal health. Until 1994 international trade in animals and animal products generally followed fair principles that allowed importing countries to protect their herds from unwanted diseases. But under pressure from key meat exporters, such as Canada, Australia, France, the Netherlands, and the United States, the World Trade Organization ended that right. Developed countries can no longer refuse imports unless they cough up lengthy "scientific justifications" or convincing "risk assessments." At the same time, the Office International des Epizooties (World Organisation for Animal Health), the global body allegedly responsible for livestock health, changed its goals from keeping infectious animal diseases in check to safeguarding world trade. Its new mandate includes such politically correct nuggets as the following: "It would be irresponsible and contrary to the principles of encouraging international trade to insist on guarantees as to the absence of commonly found infections." Or sample this reassuring credo: "The importation of animals and animal products involves a degree of risk to the importing country. This risk may be presented by one or several diseases or infections."

In the mid-1990s, the OIE dismantled its database on the occurrence of animal diseases and ended regular reporting on disease introduction through animal imports. "How can importing countries decide about disease import risk," asks Kouba, "when necessary data on epizootiological situations in exporting countries are not available?"

With trade liberalization has come a corresponding gutting of public animal health services. Although more meat now makes its way around the world, governments employ fewer veterinarians and do less disease monitoring than ever before. The world's top eight meat exporters, for example, command Augean stables with 204 million cattle, 110 million pigs, and 199 million sheep. But they employ fewer than 7,000 veterinarians to keep tabs on disease. Should it be any surprise that animal infections have spread in great explosions "into so many localities at such great distances and with such high speed"? Kouba thinks not.

Consider a brief history of rabbit hemorrhagic disease (RHD). The highly contagious virus, which causes rabbits to bleed out at the mouth and anus, probably circulated in Europe in benign form for nearly 50 years. In 1984 it popped out of a German shipment of Angora rabbits to China, where it killed millions of rabbits; many screamed in pain as they died. A Chinese shipment of frozen rabbit carcasses then took the plague to Italy, where it dispatched millions more. Another meat shipment transported the virus to Mexico, where a grocery packager innocently took the virus home to his rabbitry. The Mexicans, who had recently become fanciers of European rabbits after a persuasive advertising campaign, euthanized 100,000 creatures to control the outbreak. Since then contaminated meat or infected rabbits have bounced the plague around Europe, decimated rabbitries in England, and spilled over to wild rabbits in Spain, where scientists found dead rabbits "present in such abundance that many were left untouched by scavengers."

The Australians, who regard the European rabbit as an invasive pest and ecosystem destroyer, noticed RHD's 50 percent mortality rates. In

1995, during a sloppy biocontrol experiment, researchers let the virus escape from a facility off the coast of Australia. It then leapt onto the mainland and started eating through that continent's feral rabbit invaders with the help of insect carriers, such as fleas and blowflies. Thanks to the new global mobility of rabbit meat, fur, and pets, RHD has since shown up in Fiji, Cuba, Russia, and the Middle East.

Trade has shuttled all sorts of invasive pathogens around the world, and factory farming has helped incubate and concentrate them. Farmers once raised livestock (or what some biologists call "minion biota") in small herds that produced manure for crops and grazed local lands. Animals provided not just meat but transportation, medicines, clothing, and financial security. But in the last 30 years, farmers have been replaced by technicians, farms by factories, and livestock by "animal units." About 70 percent of the world's chickens and nearly half of the world's pigs and cattle are now raised in factories totally separate from the landscape. (Before we malled humans, we malled our livestock.) But a crowd of livestock under one roof is to micropredators what a cheap buffet is to North Americans: easy and gluttonous dinning.

The tools of factory farming combined with intense cereal production have resulted in an explosion of livestock numbers, or what many experts call the "livestock revolution." Since 1961 the number of chickens, ducks, and turkeys on the planet has quadrupled from 4.2 billion to 15.7 billion. The number of cattle has climbed from 410 million in 1890 to more than 1 billion. Pig totals have exploded from 177 million to more than 1 billion too in the last century. If you stacked all these animals on a scale, they would add up to twice the weight of all humans on earth. This revolution, in turn, has made pharmaceutical companies the fastest-growing segment of the global economy. Mountains of meat require mountains of antibiotics (40 percent of this pharmacopoeia goes to livestock), steroids, and vaccines.

All of this growth has come at the expense of genetic diversity. About 15 species of goats, cattle, pigs, and fowl account for 90 percent of what gets thrown on the global barbecue. In the last 8,000 years, farmers have bred these animals to resist disease, endure drought, and survive cold. The Navajo, for example, nurtured sheep that didn't mind desert heat, could graze anywhere, and produced a fine wool. Folks in Madagascar carefully perfected the Renitelo cow, an animal particularly well adapted to the island's climate. In the 19th century, farmers in the United States favored the American mulefoot hog because it was almost completely resistant to most pig diseases. Some stock in Africa grew resistant to sleeping sickness and heartwater fever. Scotland produced a hardy sheep that could eat seaweed, and Brazil's Crioulo Lageano cattle thrived without feed supplements and lived long lives.

But factory farming, a genuine North American innovation, has actively worked to replace these diverse breeds with specialized animals selectively bred to do one thing: grow lots of meat fast. In the trade these meat machines are known as "high-output, low-input" creatures. Today 80 percent of all dairy cows, for example, are Herefords, and most egg-laying chickens are white leghorns. Only four breeds of pigs now account for the bulk of the world's bacon. The Angora rabbit counts for most of China's rabbit meat trade.

All around the world, dumb meat machines raised in climate-controlled buildings are now replacing smart livestock that didn't need vaccines, antibiotics, or formulated rations to live well. Hardy chicken breeds that both laid eggs and made a good soup have disappeared. According to the United Nations, about 1,350 of the world's remaining 6,500 livestock breeds now face extinction. Every week an average of two creatures, representing hundreds of years of careful breeding for climate adaptation and disease resistance, disappear. The loss is irreversible. Ten percent of Asia's breeds (everything from buffaloes to partridges) are at risk. Half of all the livestock breeds that Europe depended on in the 19th century are gone, and of those remaining, more than 40 percent are endangered. Of 259 breeds in North America, 35 percent face extinction.

Although everyone knows that genetic diversity makes the best insurance against famine, drought, and epidemics, global trade and bad public policy have canceled that wisdom. Thanks to the marvels of concentration, the world now depends on six corporate chicken breeders and two corporate turkey breeders. Clem Tisdell, an Australian economist, calculates that declining domestic animal diversity not only has created a disease paradise for invaders but also might well "constitute a time bomb for the collapse of livestock production." Kim Cheng, a professor of animal science at the University of British Columbia, adds that loss of animal genetic diversity is an invitation to "a disaster that will make the bird flu situation pale in comparison."

Although the media don't care much about livestock time bombs, such disasters have a nasty habit of turning civilized societies upside down, with all the anguish of a medieval plague or 19th-century cholera. Avian flu is now on that disturbing track among Asian farmers; foot-and-mouth left a bitter political legacy in England. But to really appreciate the dangerous wildfires being kindled by international trade and crowded livestock factories, meat eaters should ponder the Masai and their calamitous encounter with rinderpest in the 1890s. This 19th-century globetrotter created a hellish tribulation now known as the Great African Pandemic.

At the time, the Masai and other pastoral tribes in East Africa depended on cattle the way North Americans today depend on supermarkets for food. The Masai viewed the cow as the perfect God-made creature because it gave milk, meat, comfort, and ease. They drank cattle blood mixed with milk and traded the beasts for iron, Chagga spears, and grain. They slept on cattle hides and healed wounds with cattle urine. They sang sweet songs to cattle and decorated the animals with beautiful tattoos. Because they believed an ancient rain god had given them the right to care for the world's cattle, they regarded cattle theft as a blessed activity. To the Masai, cattle represented the highest currency, and every dowry included a bevy of cattle.

Like other skilled husbandmen in East Africa, the Masai even used cattle to manage disease. In particular, they adopted grazing practices that kept the tsetse fly in check. This bloodsucking insect, which has been around for 34 million years, carries a parasite that eats up cattle and cattle herders alike with fatal sleeping sickness. The cattlemen knew that the fly preferred bushy country, where wildlife such as antelopes and buffaloes often carried it. So they isolated the fly belts from grazing lands with regular bushfires and intense grazing. They also avoided fly-infested zones by moving stock at night. Because the fly abhors both cattle and human excrement, herders smeared their animals with a combination of dung and milk. Long before the Europeans arrived, the region's cattle culture kept the tsetse and the parasites carrying sleeping sickness under control with what one ecologist later dubbed "an agro-horticultural prophylaxis." Rinderpest, however, changed all that.

Before its awful African debut, rinderpest (German for "horned cattle plague") had already made itself a formidable reputation as a costly trade pest and livestock killer. Although the average beef eater has never heard of the virus, scientists (and bioterror experts) still think of it as the world's most damaging virus infection affecting farm livestock. The virus, the parent of human measles, emerged some 8,000 years ago in Asia and was later called steppe murrain or cattle plague by Europeans, who noted that the contagion almost always arrived from the east. The murrain started with high fevers and ulcerated gums and ended with diarrhea and death about 12 days later. Spread by direct contact or water fouled by cattle dung, the invader often buried up to 80 percent of a herd and precipitated many a famine. The lively trade in gray steppe oxen, a boisterous draft animal but a silent carrier and shedder of the virus, started a string of epidemics several hundred years ago. According to the University of Edinburgh veterinarian Gordon Scott, rinderpest left bucolic European villages with "no transport, untilled fields, starving peasants and overthrown governments."

After one horrific epidemic in the 18th century that decimated live-stock around Rome, Pope Clement XI instructed his physician to find some solutions. Giovanni Maria Lancisi obliged and concluded that the best response was restricted trade, regular meat inspections, and burial of infected animals in lime. He also recommended controlled slaughter (mass clubbing, to be precise) to reduce spread of the disease. Cattle traders who didn't obey the rules were hanged, drawn, and quartered; priests and monks who didn't follow the rules were sent to the galleys. This unpopular inducement to obedience probably helped to restore health to the pope's cattle herds in short order. To this day Lancisi's edicts remain the bloody and brutal foundation for controlling disease outbreaks in livestock (minus the quartering of farmers, of course). Veterinarians call it "stamping out."

During repeated outbreaks in the 18th century, rinderpest killed more than 200 million animals throughout Europe and starved many a peasant family. In response governments belatedly established veterinary schools in France, Holland, and Germany to help control cattle plague and other diseases. But the virus kept moving through Europe's exten-sive trade network along with its oxen. Steamship lines took the virus for a global tour in the 19th century. Dutch and American traders quickly brought the invading virus to Indonesia and the Philippines in the 1870s.

The pandemic that changed the map and flora of East Africa began in 1887. That's when Italy's colonial army imported infected cattle from India to provision troops invading Abyssinia (called Ethiopia today). (Invaders always work together.) Rinderpest, a true newcomer, walloped the region with a wave of ecological changes that undid entire civilizations.

The invader first dispatched about 95 percent of Ethiopia's cattle. Then drought and locusts consumed local crops, with the result that entire villages of starved peasants became skimpy meals for wolves and hyenas. (All in all, a third of Ethiopia's people succumbed to famine between 1888 and 1892.) The virus then burned its way through Somalia and Sudan. In 1890 it started to dine on cattle in Masailand (Kenya) and spread down the coast to the Cape of Good Hope and German Southwest Africa (Namibia). The undiscriminating virus also started to gobble up wild ungulates, such as wildebeests and antelopes, by the thousands. In lands where cattlemen once cared for herds of 40,000 cattle, fewer than a hundred head survived the rinderpest epidemic. In his elegant account of the tragedy, the ecologist Helge Kjekshus calculated that 90 percent of East Africa's 5 million cattle perished. Wildlife died in equal numbers.

A French missionary in the British colony of Basutoland (Lesotho) recorded a familiar lament: "No more cattle, no more milk, what shall we eat? No more cattle, no more fuel, what shall we use for making fire? No more cattle, no more skin clothes, what shall we wear? No more cattle, no more marriages, how shall we marry? No more cattle, no more ploughing, what shall we eat and where shall we get money?"

The devastation horrified European colonialists, including Fredrick Lugard, an observant and sympathetic British administrator in Kenya. He called for imperial veterinarians to come witness the die-off, but none arrived in time. "Never before have the cattle died in such vast numbers; never before has the wild game suffered. Nearly all the buffalo and eland are gone," he reported, dumbfounded. Even giraffes succumbed to rinderpest. To the Masai, the invader appeared to be an ordained catastrophe: "First of all there was an eclipse of the sun and it took place at about 5 o'clock in the afternoon.... It was then that the rinderpest attacked the cattle. The epidemic finished the Masai cattle."

It almost finished the Masai, too. Oscar Baumann, an Austrian geographer, traveled through lands belonging to the Masai during the plague and found perdition: "skeleton-like women with the madness of starvation in their sunken eyes, children looking more like frogs than human beings, 'warriors' who could hardly crawl on all fours and apathetic, languishing elders." The survivors even ate dead donkeys and cattle horns. Two-thirds of the Masai people died of starvation. Another observer reported that the corpses of cattle and cattlemen were so thick on the ground that vultures "forgot how to fly."

The livestock plague changed the look of East Africa forever. Europeans scooped up abandoned real estate and concluded that the country was fairly empty of people. Wild game populations rapidly recovered, creating the modern myth that Africa, the birthplace of *Homo sapiens,* was nothing more than a wild paradise for white big game hunters. Other biological invaders took advantage of the mayhem. Smallpox, which locals called "white illness disease," finished off many survivors of the rinderpest-induced famine. Then came the South American sand flea, introduced in the dry ballast of a British merchant ship. Sand flea larvae burrow into people's feet. Left unattended, they can grow to the size of a pea and cause blood poisoning. Africans had seen nothing like them and did not know how to respond. The ravages of the fleas left thousands of Africans deformed, with only one or two toes or a shapeless stump.

After the flea came the fly. With cattle herders and their prized grazing animals gone, grasslands grew into bushy woodlands, the favored terrain of the tsetse. The parasite did what any self-serving biological invader would do: It expanded its territory. Between 1900 and 1906, more than 200,000 East Africans died of sleeping sickness, a scourge that their cattle culture had previously held back. Europeans called the epidemics "mysterious" and labeled the country unhealthy. They also converted depopulated lands into parks and game reserves, such as Serengeti, Selous, and Tsavo, entrenching the tsetse fly's contribution to the stereotype of wild Africa. The pest still bedevils 9 million square kilometers (3.5 million square miles) of land in 37 countries.

Rinderpest left other troubling legacies that might even astound Stephen King. After the alien virus ate up cattle herds and the region's wild buffalo, East Africa's famished lions went on several notable rampages. The famous man-eaters of Tsavo, for example, put a stop to the construction of the Uganda Railway in 1898 by consuming 28 Indian porters.

Later outbreaks of rinderpest in Africa also put more humans on the lion's menu. When the virus again decimated Ankole cattle herds in Uganda in the 1920s, the colonial authorities decided to wipe out all the wild ungulates in the region, in the hope of ending the virus's reign of terror in domestic herds. Unfortunately, they succeeded only in eliminating the local food base for big cats. Hungry felines predictably switched to two-legged dinners. One man-eater eventually dispatched 84 people; another achieved a more modest toll of 40 Africans. Their predations proved so fierce that pastoralists abandoned entire villages.

Another program to halt a rinderpest outbreak in the 1940s again called for the killing of wild game in Zambia. It, too, had predictable results. Deprived of buffalo and other wild fare, lions ate more than 1,500 people over 15 years. So the bold predations of an invading micropredator changed an entire landscape, created a chain of bad decisions, and eventually invited macropredators to run amok. And some say nonfiction is boring.

Rinderpest rarely appears in African history books, but it changed society so completely that the biological clock has never turned back. (Africa gave Europe a taste of its own medicine in 1922 when an imported zebu cow brought rinderpest to Belgium and created such a trade scare that Europeans formed the Office International des Epizooties.) Rinderpest remains a curse in East Africa and a bane to wildlife despite multimillion-dollar campaigns to stamp it out.

Just as the Masai still talk about rinderpest, the British still bemoan foot-and-mouth disease (FMD). In 2001 an FMD epidemic overwhelmed the island and once again demonstrated how effortlessly international trade liberalization can spread disease. The invasion plunged rural Britain into chaos, pummeled the tourist industry, and ended with one of the greatest mass killings of animals ever witnessed by modern man. The world's foremost authority on the virus, the late Fred Brown, decried the episode as a "disgrace to humanity."

Although government officials, factory farmers, and globalists reluctantly justified all the bloodletting on the grounds that they were making the world safer for global trade, the truth is much uglier. What the British government later called a "minor victory" was nothing more than grotesque folly. Despite decades of "theoretical" preparation, the United Kingdom was about as ready for FMD as the Masai were for rinderpest and Asian farmers were for avian flu. The invasion raises a profound question: If a society as advanced as Britain can't respond to a global but nonlethal trade disease or a simple case of disease pollution without resorting to mass slaughter, how can the rest of the world ward off a genuine disease threat?

Jonathan Miller thinks a lot about that question. After leaving Yugoslavia's killing fields in the 1990s, the British war correspondent decided to find a more congenial way to make a living. So he took up a new career as an "apprentice peasant." He sold his home in London and purchased a 17th-century farmhouse in Surrey, a John Deere 6200 tractor, some Jacob sheep, and a few horses. But before he could appreciate the song of the nightingale and the peaceful life of a hobby-farm columnist for *The Sunday Times,* foot-and-mouth disease entered his backyard.

To his surprise and total consternation, Miller found himself once again in a war zone. It wasn't ethnic cleansing but ovine cleansing, and it came "complete with talk of taking out the enemy." As the government response to a rather harmless invader turned England into Plague Island, Miller recorded more "lies, spin, incompetence, cruelty, and waste" than he had ever seen on a human battlefield. To top it off, the whole catastrophe ended with "the most expensive barbecue in history."

Foot-and-mouth disease (not to be confused with foot in mouth, a largely political syndrome) is an ancient affliction that has dogged cattle, sheep, and pigs for centuries. Its original home may well be the Middle East or India, where the cow is holy and lives in unimaginable numbers. Until the 19th century, farmers regarded FMD as a cursed yet curable nuisance. But thanks to global trade and other innovations such as increasing livestock densities, says the courageous British historian and veterinarian Abigail Woods, it became a full-fledged "state-managed animal plague" of irrational proportions. It is a biological invader that simply robs the pocketbook as opposed to filling the morgue.

Unlike avian flu and many livestock diseases, FMD poses no threat to food safety or human health. The agent, the first virus ever identified in animals, comes in seven strains and is not so much a potent killer as it is an efficient misery maker. An infection produces a high fever and numerous blisters on an animal's feet and in its mouth. Afflicted animals often stop eating and go lame because their mouths and feet are so sore. Pregnant animals tend to abort and their milk dries up. Although the fever carries off weak and younger animals (the mortality rate is about 1 percent), most others regain their health within 15 days if the blisters don't get infected by bacterial opportunists. After a visit from FMD, the meat and milk production of a herd can decline by 15 to 20 percent— an unacceptable trend for the international meat trade.

Because only 1 to 10 particles are required to infect an animal, scientists regard the organism that causes FMD as the most contagious mammal virus on the planet. Carried by wind, birds, vehicles, straw, buckets, meat products (it can survive in frozen bone marrow for months), manure, semen, flies, ticks, water, and soil, the FMD virus can travel faster than Superman. Between 1957 and 1960, Vaclav Kouba documented the spread of FMD by meat products, faulty vaccine production, kitchen wastes, animal imports, mobile farmers, and portable equipment.

FMD's remarkable stickiness has obsessed bioweaponeers in Russia, the United States, the United Kingdom, Canada, and Japan for nearly 50 years. They quickly realized that just a few virus particles conveniently dropped in the right place could cause big trouble. An outbreak could destabilize and reduce a nation's meat and milk production so quickly that political chaos and even famine could result. The U.S. government's biowarfare program saluted FMD's trade-killing powers and even calculated how much would be required to deplete food supplies in the Soviet Union and China. It conducted tests on its own stockyards with a harmless simulant in the 1950s.

Governments now consider the virus so "hot" that only a few seemingly secure high-containment laboratories can handle it. The two most prominent labs (both former biowarfare facilities) are Pirbright in Surrey, England, and Plum Island near New York City. Although the United States records its last "official" outbreak of FMD as having occurred in 1929, that's not correct. It really happened in 1978, when a leak at Plum Island, the "world's safest lab," killed scores of test animals. Had the virus escaped beyond the island, it would have tanked the $100-billion livestock industry, shut down meat exports, and exploded the rural economy of North America. (The British researcher A.I. Donaldson reported in *The Veterinary Record* that "residual live virus in vaccines which had not been fully inactivated and escapes from laboratories working with the virus" accounted for 13 FMD outbreaks in Europe before 1992.)

The FMD virus first appeared in the 16th century in Italy, then a big global trader. Until the 20th century, most farmers regarded the viral invader as a mild occupational hazard. Whenever the disease struck, farmers typically displayed the head of a dead sheep or cow at the manor gate to keep drovers and visitors away. Then they gave their remaining animals meals of warm mash and soft hay, and gave them fresh bedding. They often tarred sick animals' hooves to prevent them from licking the sores and treated the blisters with tamarind, chilies, or macerated neem leaves. With good stockkeeping, the animals recovered and farm life went on. Many previously infected animals even won prizes for milk and meat production.

But a combination of political, economic, and trade interests in Britain turned the virus into what Abigail Woods has poignantly dubbed a "manufactured plague." In particular, influential breeders, often MPs and members of the Royal Agricultural Society, noticed that purebred cattle suffered great losses from FMD while routine meat- and milk-producing varieties did not. Arguing that FMD caused extra feed consumption, lowered milk production, and lengthened the time of fattening for highly genetically uniform pedigree herds, breeders persuaded the British government to make FMD a notifiable invader in 1871.

The slaughter of animals to control outbreaks of FMD soon became a disagreeable pastime in the United Kingdom as well as a "feature of British natural superiority" and national pride, says Woods. Despite great opposition from farmers and medical doctors, who regarded the practice as barbaric and unscientific, the killing persisted and the country kept its "disease-free" status.

Authorities repeatedly ignored the irrational nature of the culls. In the 1920s, a slaughter team arrived at one farm after a group of animals had recovered from FMD. The farmers pointed to their now normal cows and asked, "Was that it? Was that trivial illness what all the fuss was about?" The government killed the recuperating cattle anyway.

By the 1940s, Britain's bloody stamping-out policy had transformed FMD into a source of dread around the world. The United States adopted similar tactics that scared the hell out of livestock farmers. When the virus appeared in Mexico in 1948, threatening the Americans' disease-free status, the U.S. agriculture department literally invaded the Mexican countryside with 10,000 men trained to slaughter cattle in vast numbers. Until a vaccine could be produced in large enough quantities, the authorities decided to kill both infected and healthy cattle in order to put a "fire break" around the outbreak. Mexican peasants and even priests called the Americans *los matavacas,* the cow killers.

The resentment quickly turned bloody. In one town angry campesinos seized an American veterinarian and stabbed him to death.

For good measure they also gouged out his eyeballs. As many as 24 Americans and several hundred Mexicans died in machete clashes during the viral campaign. When the Americans finally switched from killing to vaccinating, villagers remained equally distrustful. Some believed that eating meat from vaccinated animals might sterilize them or that the campaign was all part of a sinister U.S. plot to depopulate Mexico. In the end the Yankees stamped out *aftosa* at the border but left behind a trail of resentment.

FMD didn't acquire its modern reputation as a vicious, rampaging virus until 1967. That's when an imported lamb from Argentina (where the virus had been introduced by British trade in the 19th century) probably shed the pathogen on a pig farm on the Cheshire plain. The region, an early example of the Guangdong syndrome in terms of animal density, then had one of the highest concentrations of livestock in the world. Starlings and gulls likely spread the virus from there, and economic chaos followed. Farm markets closed, trade stopped, and much of central England was quarantined. The government called out the military to decontaminate farms, set up roadblocks, and help slaughter half a million animals.

One funeral pyre alone required 450 bales of straw, three truck-loads of tires, 900 liters (200 gallons) of oil, 250 railway sleepers, and 40 tonnes (45 tons) of coal. Clouds of smoke from the fires canceled plane flights. Within six months, however, the government had contained the epidemic and resumed trade with Argentina, against the protests of local farmers. "The world goes around again," concluded Ralph Whitlock, one astute chronicler, "and without much doubt in its circling it will bring yet another installment of the Great Cattle Plague."

In the last decade, the global meat trade has recirculated FMD with a vengeance. As the world's livestock herds grew by 60 percent from the 1960s to the 1990s, FMD simply expanded its range. The invader spilled over from crowded pens of livestock in Asia and Africa to

European factories. A new virulent strain emerged in India in 1990 and spread from Saudi Arabia to Turkey as well as Nepal and China. In 1997 it arrived in Taiwan in a batch of illegally imported feed or livestock. There it crushed the nation's celebrated pork economy as only livestock plagues can do.

At the time the island boasted one of the world's highest pork densities (more than 11 million hogs) as well as the world's highest pork consumption (40 kilograms, close to 90 pounds, per year per person). More than 1 million people worked in Taiwan's industry raising pigs, growing feed, or freezing bacon. The island had become so besotted with hog profits that every waterway had been fouled by hog manure. But pork prices never reflected these costs. Japan, which bought 40 percent of the island's production, about $1.6 billion worth of meat a year, just wanted cheap chops.

The outbreak changed all that, as the virus moved from 28 farms to 217 farms in just one week. Although the disease killed only 100,000 animals, hundreds of distressed soldiers armed with electronic zappers slaughtered an additional 4 million pigs as a precautionary measure to keep the virus from spreading. The government voted to vaccinate the rest because it simply didn't have the resources (or the British resolve) to slaughter 6 million animals. Japan cancelled its pork orders anyway, as did the rest of the world. During the culling and vaccination campaign, the Dalai Lama said a prayer for Taiwan's pigs and wished them well in their heavenly recycling. Buddha himself might have offered a more critical homily: "Human greed and foolishness are like a fever. If a man gets this fever, even if he lies in a comfortable house, he will suffer and be tormented by sleeplessness." The epidemic had one good effect: It reduced the island's pork industry to a more manageable size.

In 2001 the next bout of fever hit Britain, where the virus found everything it needed to cause much sleeplessness. Ideal conditions included unprecedented rains (like avian flu, FMD favors moist conditions), rising volumes of trade (the United Kingdom exports a million live lambs to the Continent every year) coupled with reduced border controls, a decimated government veterinary service, and unprecedented pig and sheep densities. As one pig farmer put it: "Supermarket greed and the drive for globalization at all costs has turned this country into a cesspit for the world's cheapest meat and meat products."

According to official records, the epidemic probably started on Burnside Farm at Heddon-on-the-Wall in northern England. There Robert Waugh, a 55-year-old swineherd, admitted that he fed his animals swill made from restaurant scraps from the local airport or other establishments. The feed, both cooked and uncooked, probably contained meat infected with FMD from Asia. But the Ministry of Agriculture, Fisheries and Food (MAFF) had also been secretly experimenting with a new FMD vaccine in the northeast, and many suspect a leak occurred. Waugh, however, was one hell of a messy pig farmer (visitors found dead stock feet up in piles of manure) and conveniently became the government's scapegoat. But thanks to the volume of illegal and legal meat trade, it's most likely that a mild strain of the virus had been moving around the country long before it undid Waugh's farm and ended his farming career.

Every year 29,000 tonnes (32,000 tons) of illegal mutton, beef, and pork enters Britain. (Epidemiologists recently calculated that probably 100 tonnes [110 tons] are infected with FMD.) According to the UN Food and Agriculture Organization, dirty or black-market meat likely caused 7 out of 11 primary outbreaks of FMD in Europe from 1991 to 1996. The traffic includes all sorts of delicacies, such as cattle feet from France, bush meat from Africa, and sheep carcasses from the Middle East. The Nigerians particularly favor mutton blowtorched black for its distinctive scorched flavor.

From Burnside Farm the virus spread quickly through England's huge sheep population (31 million, up 80 percent since 1978), because of the miserable distances modern livestock must now travel to a slaughterhouse. Since the 1967 FMD outbreak, which had been contained in a local area with few roads, the livestock industry had grown more concentrated. Fewer farmers operated bigger facilities containing larger flocks and herds and increasingly relied on truck transport to reach fewer markets and retailers. Where there had once been 800 markets, now only 170 stood. Some 3,000 abattoirs had now become 520.

As a consequence livestock were moving around England like crowds of morning commuters. A single stock dealer, who specialized in moving sheep from here to there so farmers could qualify for extra European Union agricultural subsidies, introduced the virus to Scotland, Northern Ireland, and the Republic of Ireland. Before the British government realized what had happened, a mild, almost symptomless infection had popped up in more than 100 farms and markets.

As *The Guardian* and other newspapers reported, the government did its best to turn a crisis into a national disaster. First it failed to impose an immediate ban on animal movement after FMD's initial discovery in February, a delay that doubled the number of creatures that the government eventually slaughtered. Next it dusted off its

Paul Faith/PA/Empies.

Britain's 2001 foot-and-mouth outbreak resulted in a Rwanda-like slaughter of sheep: the biggest cull of livestock anywhere on earth.

19th-century stamping-out policy. Stamping out can work at the beginning of an epidemic. But the government had already missed that important appointment with the virus by three weeks.

It also ignored the advice of the world's foremost animal virologist and FMD expert, Fred Brown. The biologist, who spent his life studying the pathogen and its genetic makeup, quickly suggested that a "ring vaccination" campaign around infected herds would likely quell the epidemic, lessen viral production, and spare the lives of harmless animals. He also offered a simple blood test to the government and noted that previous vaccination campaigns on the Continent had stopped the invasion. In fact, most of the world was already using vaccination to control FMD.

Instead of following Brown's sound advice, agricultural technicians dragged out an outdated disease control plan that showed 800 farms as having "zero animals" when, in reality, all contained animals infected by the epidemic. Next they plugged in an unvalidated computer model designed in New Zealand that completely underestimated the spread of the virus. (During the epidemic, techies routinely sent data on the invader's progress to New Zealand, only to wait five or six hours for a reply.) The urban professionals using the model ("a tool to calculate virus production on each farm") had no idea how factory farming worked and knew even less about the movement of livestock. Within five weeks the epidemic was completely out of control. Because of a critical shortage of veterinarians, the government solicited contractors to slaughter and burn animals. The skies began to go dark throughout rural Britain.

After five weeks of mayhem, black smoke, and misinformation, the Cabinet Office Briefing Room (COBR), a special committee designed to deal with terrorist threats, took charge of the epidemic on March 21. It met in a subterranean bunker equipped with TV screens. As two crusty law professors later noted, a regulatory problem of livestock rearing and farm economics had passed to "government apparatus designed to deal with problems more akin to general insurrection ... with little more than approving comment in official reports."

The Orwellian briefing room ignored pleas to "vaccinate, not eradicate" and immediately embraced a 24-hour/48-hour slaughter policy championed by the computer modelers. All infected animals (and one in three diagnoses was incorrect, because of more bungling) had to be killed within 24 hours, and all hoofed animals within a 3-kilometer (5-mile) radius of "the zone of the infected" were to be exterminated within 48 hours. To the professionals sitting in front of screens in London, the extermination plan looked like a smart video game. But to livestock owners, it quickly became "carnage by computer." The cull ignored geographical barriers to the virus, the breeds of animals, the importance of genetic diversity, the type of farming practiced, or even the susceptibility of the livestock in question.

The brutality and draconian nature of the extermination horrified rural Britain. Lambs shot by a captive-bolt gun did not die; the gun simply stunned them by compressing their skulls or blowing away their jaws. "This can't be right," wailed a farm wife. "My husband is waiting to send 1,500 healthy sheep to be murdered." In one case, three carloads of police and one vet showed up to kill one pet goat that had grazed next to a farm simply suspected of infection. Children who questioned the killings were told by officials, "Grow up. This is the real world, not Disneyland."

One slaughterman contracted by MAFF kept a diary of the killing. In March 2001 he wrote: "120 cattle slaughtered this morning. What a sight. They shot a bull and he just would not die. He was shot four times and pithed everytime. He just would not go." (Pithing means severing an animal's spinal cord, usually with a knife.) In April he watched at one farm where "sheep still not dying fast enough, slaughtermen moving so fast through the pen that shot sheep are trying to get back on their feet." On another day a cow tried to defy the killing squads: "As they tried to pith her, she stood up and ran round the barn. Six slaughtermen jumped on her and shot her twice more. She must have been demented with fear. Every slaughter and welfare rule is broken here and MAFF just stand by and watches…. Had enough of this madness and called it a day."

The quarantines made rural folk prisoners in their homes. They could not attend funerals or weddings, and children did not attend school. And when they did move about, they often drove past piles of burning animals. A community nurse, with two young girls in the car, took a wrong turn and found a perversion of agriculture: "By mistake I took them through a closed road, a sign having fallen down to the side of the grass. In the dark we went past a burning pyre only yards from the hedge separating the road and the field. We could see the charred, rigid bones of the cows and sparks from the fire and the smell permeating the air and the silence of the two young girls."

The invasive butchery appalled many veterinarians employed by the government. Forty wrote a damning letter to Prime Minister Tony Blair, pointing out that the FMD command center was being led by men with no veterinary training. "As a result we are seeing a savage attack on what livestock remains in the north of England and the southwest of Scotland. Animals are being slaughtered without rhyme or reason, often weeks after the supposed danger farm has been eliminated. This 'scorched earth' policy will undoubtedly result in the eradication of foot and mouth disease, but it may be a pyrrhic victory."

By now thousands of animals lay in heaps by farmhouses, where reporters found them "unburied, stinking and oozing fluids into the groundwater." Prized flocks of bison, alpacas, goats, and hefted sheep were left unattended in the backcountry while packs of foxes gorged on the carcasses of the slaughtered beasts. Government incompetence continued to escalate faster than piles of dead sheep. Throughout Britain contract killers got their directions wrong and showed up at the incorrect farms but started shooting animals anyway. Those animals not killed with a bullet were either pithed or injected or had their throats cut. Inexperienced vets often broke off four needles while attempting lethal injections in an animal's chest before finding the heart. Some days as many as 50,000 terrified animals in crowded lots, short of water and feed, were butchered the same way Nazis rounded up and shot peasants in Eastern Europe during World War II. One former slaughterman characterized the killing as too cruel for belief: "If some of these so-called slaughtermen were killing my stock, I know where I would be pointing the gun, and it wouldn't be at the sheep."

The panic and quarantines prevented farmers from transporting feed; many became unable to feed their own animals. So the government started killing animals for "welfare" reasons. It rubbed out another 2 million animals to prevent cannibalism (a real problem in pig factories) or death by starvation. In the end the government exterminated and burned 10 million animals on 10,000 farms. By most estimates fewer than 10 percent of the slaughtered were actually infected.

Overwhelmed by the logistics of the largest slaughter of animals in British history, the government belatedly called out thousands of soldiers and Royal Marines. They set up quarantines and roadblocks, and reluctantly helped with the cull. Brigadier Alex Birtwistle couldn't believe the chaos or the "startling level of incompetence at every level." He found rotting heaps of carcasses outside of people's homes and animals so decayed that the carcasses fell apart while being moved to pyres. "There was no contingency planning at all."

The scale of the holocaust might have boggled even the pest-ridden imaginations of medieval plague victims. Some 15,000 vehicles trundled through the countryside, splattering the roads with blood as the heads and legs of sheep and cattle bounced up and down. The tonnage of dead stock moved each week surpassed the weight of all the ammunition transported by the British armed services during the Gulf War. The engineering effort, which required more than 10,000 people, involved excavations equal in size to 200 Olympic-size pools. One mass burial accommodated 430,000 sheep.

The killings blackened rural life and devastated the tourist economy. (Tourism was never part of the virus computer model.) To this day some economists still refer to the epidemic as the worst social and economic catastrophe to befall Britain in decades. In rural areas business dried up and the pubs and hotels sat empty. A dead silence hung over the countryside. Parks were closed and hunting and fishing stopped, as did most sport events. The tourist industry lost $100 million a week because foreign visitors didn't want to see a Rwanda for animals or be blinded by smoke. Grown men cried and more than 60 farmers committed suicide. Prime Minister Tony Blair even postponed an election.

In all, the government spent more than $20 billion to control a viral invader that posed no threat to food safety or human health, for an industry that generated only $653 million in meat sales a year—and all to protect Britain's "disease-free status." But the interests of global trade prevailed, and Britain's livestock industry emerged more concentrated and industrialized than ever before. Since the outbreak more than 10,000 farmers have left farming, thanks to FMD.

The human cost continues to haunt Britain's rural residents. A study by the Institute for Health Research at Lancaster University that looked at the weekly diaries of 54 survivors of the disaster found a grim legacy. "Life after the foot and mouth disease epidemic was accompanied by distress, feelings of bereavement, fear of a new disaster, loss of trust in authority and systems of control, and the undermining of the value of local knowledge."

Biowarfare experts and the U.S. Marine Corps studied Britain's FMD horror show and came up with explosive conclusions. The Marines, for example, calculated that any similar outbreak in North America would seriously affect national security because a large percentage of army installations happen to be located in agricultural areas, such as North Carolina. A livestock plague could result in a lockdown, confining military movements to disease fighting or to bases. In addition, there would be widespread disruption of interstate and intrastate commerce and transport. Tens of thousands of soldiers would be needed to contain the outbreak. Factory farming combined with a livestock plague could, in fact, "have serious impacts on military operations."

Peter Chalk of the Rand Institute commented that FMD had made science fiction a reality in Britain. A simple virus that was so transmissible that no one needed to weaponize it could defeat a country without killing even one person. The dictates of global trade had created the equivalent of an act of bioterrorism: "You've got economic impact, you've got destabilizing of the government, you've got social attacks and you've even got the possibility of mass scare." Just imagine, added Chalk, what might happen if you had a disease that was transmissible from animal to human.

In 2002, a year after the outbreak, an exhaustive report by the European Parliament lambasted the British government and the OIE for outright deficiency and sheer idiocy. It blamed the epidemic squarely on shocking livestock densities as well as increased trade and animal movements. It noted that global trade had accelerated the spread of diseases without "any corresponding expansion of inspections or veterinary system, research into vaccines and modern testing equipment." In other words, global traders had exposed agriculture to more disease at the same time that governments had dismantled infrastructure to deal with invasive species, because "there is a general neglect of biosecurity issues when driving trade liberalization measures forward."

The report also condemned special interest groups, particularly food traders and some farm lobbies, that had supported the unprecedented slaughter instead of a vaccination campaign simply to avoid livestock trade restrictions. And it characterized Britain's unvalidated computer model as junk that yielded a strategy "fraught with inevitable lax biosecurity and documented infringements of animal welfare law." In a separate report, Vaclav Kouba noted that "the preparation of veterinary manpower for dangerous disease prevention, diagnosis, control and eradication under an emergency situation has been in general … a total failure." Kouba also blamed the OIE for "favoring disease export regardless of animal and human health consequences" and consistently violating the ancient medical code of *Primum non nocere* (First do no harm).

Christopher Stockdale, a farmer traumatized by the shooting of his healthy livestock during the epidemic, wrote a lively dissertation on the outbreak in 2004. His primary conclusion: "Progressive socio-political de-prioritisation of agriculture coupled to political demand for ever-cheaper food were likely to combine at some point in such a way, or worse, and may again."

In an edgy essay titled "Carnage by Computer," two Welsh lawyers, David Campbell and Robert Lee, noted that the FMD epidemic "involved lawless action by government on such a scale as to amount to a negation of the basic precepts of the rule of law." They bluntly added that government agricultural policies grossly ignored the risks of increasing herd size and animal traffic while actively dismantling disease control systems. "If a private business stored such a large quantity of flammable materials on its premises that its fire control measures could not cope with the great size of a fire caused thereby, would it be excused from liability for the damage caused, even if it honestly pleaded in its defence that it hadn't kept an accurate estimate of the risk because it had no idea what quantity of materials were stored?"

Although it's likely that no government will ever respond to an FMD invasion with mass slaughter again, vaccination alone won't stop the meltdown now being experienced by the world's livestock. In the last two years alone, FMD has unsettled rural lives in Brazil, Peru, Israel, Botswana, Colombia, and Hong Kong. Until industry and government reduce the fuel for livestock invaders—namely, overcrowded factories and the mass movement of live animals—more pathogenic fires will burn brightly in the countryside. So the livestock revolution will continue to provoke a biological counterrevolution with obscene costs for taxpayers, mass death for animals, and someday a Masai-like hell for us all. "We are headed for another disaster," warn Campbell and Lee, and that's probably an understatement.

Before Fred Brown passed away in 2004, the eminent virologist reflected on foot-and-mouth's carnage in Britain with a rare honesty that someone, somewhere, might someday apply to avian flu: "As time goes on scientists know more and more about less and less while the politicians know less and less about bugger-all."

FOUR

The Triumphant Prion

*"For it has been a matter of common observation from
the earliest times, and our history will testify to its accuracy,
that widespread pestilence in plants, and murrain in animals,
have frequently either preceded, accompanied or followed
closely on those visitations which caused mortality
and mourning in the habitations of men."*

—George Fleming (1871)

Maurice Callaghan, a mechanical engineer, father, and avid cyclist, had two pieces of bad luck. He ate beef three times a week, and he shared a genetic weakness found in only 39 percent of the British population. That anomaly left him completely defenseless against variant Creutzfeldt-Jakob disease, or what doctors call the human version of mad cow.

Callaghan's maddening symptoms appeared in February 1995, when the British government still swore that beef was safe to eat. Like half the British population, Callaghan was having trouble sleeping at night. But then he fell down the stairs and his speech began to slur. He forgot his PIN bank card number and lost his abiding interest in cars. When the 30-year-old's handwriting became shaky, doctors referred him to a rheumatology clinic in May. As Callaghan's condition worsened, he often dressed and prepared to go to work at 9 p.m., confusing it with 9 a.m. Then he started to hallucinate and defecate in his clothes.

By August the man couldn't walk or talk. While his family toileted, washed, and changed him upstairs, his two daughters quietly watched videos downstairs. The presence of flies particularily rattled him; he died a shrunken ghost of a man in November. To his family's horror, medical authorities inexplicably dug Callaghan a "medieval grave" 3.2 meters (10.5 feet) deep and lined it with lime. A year later the British government belatedly admitted that eating beef could drive some people crazy.

The brain-wasting disease that killed Callaghan and 160 other Europeans as well as hundreds of thousands of cattle in the 1990s stands as one of the creepiest and most insidious invaders on the planet. Its evolution from a relatively bizarre sheep affliction to a trade-stopping global contaminant reads like an Edgar Allan Poe yarn complete with crazed mammals, beef-chomping politicians, opportunistic traders, and civil servants with the science literacy of third-graders. Mad cow disease, or bovine spongiform encephalopathy (BSE), also illustrates once again how global trade can take an ancient and obscure group of troublemakers, such as transmissible spongiform encephalopathies (TSEs), and give them, as the experts say, "spectacular geographic amplification."

Although mad cow appears to be more of a calamitous trade saboteur than an Armageddon pathogen, the scientific jury remains both confused and alarmed. Every year scientists learn something new about the behavior of TSEs, and every month the invasion of the world's remaining 4,500 mammal species by these illnesses expands exponentially with the help of barnyard cannibalism and industrial farming. During the lengthy BSE Inquiry in the United Kingdom in 1998, one of Callaghan's brothers, Gerard, angrily described variant CJD as a "frightening, rapidly destructive dementia." He added that it was a "disease whose source lies in the mistakes of the past ... a disease whose course lies in the fortunes of the future."

But first a word on prions, something completely different in the world of invaders. Most scientists now believe that TSEs are caused by a specific protein (one of the hundreds the body uses to help it think or digest) that carries no genetic material. This rogue protein, the prion, attaches itself to normal relatives in a cell and then bowls them over. Scientists call it folding. Over time the bad protein slowly works its way to the brain, where the exponential growth of misfolded proteins eventually shoots the gray matter full of holes. The late American neurologist Paul Brown compared this astounding conversion to "turning a chiffon curtain into a Venetian blind." Stanley Prusiner, who won a Nobel Prize in 1997 for describing these killer proteins (he named them prions), champions this point of view.

But this new medical orthodoxy remains a hypothesis, and many brilliant scientists, such as Laura Manuelidis, a TSE expert at Yale, still argue that prions may be a by-product or symptom of another undiagnosed invader (such as a slow virus). Everyone agrees, however, that these brain eaters defy elementary biology by mimicking much basic chemistry. For starters prions are smaller than the smallest-known virus and can survive in tissue long after death. They can cross the species barrier and incubate for decades. Incredibly, they produce no immune reaction. They appear to be almost indestructible and can contaminate pastures, animal feed, feeding equipment, medical tools, and blood. Once in the brain the agents perform like clockwork and fold healthy proteins into long hole-making chains. TSEs are invariably fatal.

In all likelihood TSEs have dogged cattle, sheep, and humans for thousands of years. But until fairly recently, brain eaters emerged only sporadically, caused isolated clusters of inexplicable madness, and then disappeared.

The triumphant march of the brain eaters began with scrapie in sheep. In the 18th century, sheep farming became Europe's livestock king and a pillar of economic well-being. About one in four Europeans put mutton on the table or spun wool for the market. That meant the population of sheep outnumbered or equaled the human populations of rural England, Germany, and France for the first time ever. In fact England had just as many people as sheep in the early 1700s: 10 million. That kind of crowding probably broke a critical threshold for the world's first-known brain eater. Somewhere along the industrial sheep line, a wooly ovine ate bits of a rendered relative, or sheep farmers unknowingly pampered a breed susceptible to a brain eater.

The first cases emerged in the 1730s and killed 10 to 30 percent of some flocks. Shepherds had seen nothing like it. According to the Reverend Thomas Comber, an afflicted ram at first looked wilder ("as though he were pursued by Dogs") and then rubbed itself on trees or posts ("with such fury as to pull off his Wool and tear away his flesh"). Sensible English farmers called the disease scrapie because of all the frenzied scratching and itching, while French farmers dubbed their invasive strain *tremblante* (the shakes).

Scrapie unsettled the sheep industry because no cure could be found. Wool was in such high demand that every animal counted. Nonetheless many farmers dealt with the invader by employing an ancient remedy: they slaughtered any mad beasts. The meat was given to the servants. The poor fed the distempered animals to their children.

Given the demands of the wool industry, scrapie soon dominated pub talk in the 18th century, much as mad cow did in the late 20th. In 1755 farmers even took their case to the House of Commons. They complained that the disease had ravaged their flocks for nearly 10 years; it seemed transmissible from adults to offspring (was "in the blood"); and it couldn't be cured. They loudly blamed its rapid spread on free traders. During sheep auctions, unscrupulous jobbers were stealthily mixing scrapie-sick sheep in among healthy flocks in order to earn maximum profit. Similar trade deceptions later transported BSE around the world.

As the wool trade waned, scrapie disappeared from political debates but continued to worry sheep farmers and veterinarians. In the 19th century a Scottish vet clearly established that the agent was indeed contagious; moreover, it had an incredibly long incubation period—18 months—and appeared only in animals two years or older.

Meanwhile the livestock industry inadvertently opened another pathway for killer proteins into the human food chain. In the 1860s, a charismatic and brilliant German chemist, Justus von Liebig, concluded from his study of plants, minerals, and other compounds that humans and other mammals worked best on a concentrated fuel of protein. To walk the talk, the chemist developed a protein-based replacement for breast milk ("Liebig's Infant Food") as well as his very own meat extract made from boiled-down beef offal. For animals (particularly pigs) he also designed in 1865 a steam-based rendering system that turned animal carcasses into pools of fat and a sugarlike powder of meat and bonemeal. Thanks to Liebig, the "scientific feeding of animals," or industrial cannibalism for ruminants, began and soon grew to a grand scale.

Pigs and chickens, both voracious omnivores, had no problem eating their ancestors. Devoted grass eaters, however—horses, oxen, and sheep—balked at the change in diet, suffered indigestion, and ate their own kind only reluctantly. But the efficiency of recycling dead ruminants, combined with cereal shortages during the two world wars, kept meat and bonemeal (MBM) powder on the livestock menu. By the 1930s most European farmers routinely fed animal protein to pigs and chickens. Agricultural chemists eventually got dairy cattle hooked on MBM, as it's known in the trade, by mixing it with well-liked cereal concentrates.

Liebig's protein revolution made Germans, voracious meat eaters, into the subjects of an uncontrolled science experiment. With the exception of a wise Canadian pathologist, Murray Waldman, no one has really explored this historical development. It is both curious and intriguing that, just 40 years after Liebig's innovations, German neurologists were the first in the world to describe cases of human dementia. Until then, medical texts rarely made mention of such disorders among the old: people just died with clear minds.

That pattern changed dramatically at the beginning of the 20th century. In 1906 Alois Alzheimer wrote about a 50-year-old matron who had lost her memory, began to scream loudly, and died four years later. He called it "an unusual disease of the cerebral cortex." Alzheimer's disease is now an epidemic that primarily plagues citizens of nations with highly industrialized animal-feeding and meat-packing systems. According to Waldman, Alzheimer's now affects 1 in 10 persons older than 65. The residents of India, who admire the cow but don't eat it, have very low rates of Alzheimer's. So, too, do aboriginals and Africans who dine on country meats. In contrast, poor Scots and poor black Americans, who eat large quantities of processed protein, have extraordinarily high rates of Alzheimer's. "The geographical evidence strongly suggests further investigation is urgently needed," adds Waldman.

But Alzheimer wasn't the only observer who spotted trouble in the brain after Liebig's protein revolution. Shortly after his discovery, two other neurologists described a slightly different kind of madness. Hans Creutzfeldt, a student of Alzheimer's, wrote about a maid who went crazy in 1913 with tremulous arm movements and "uncontrollable bursts of laughter." She died paralyzed, with her face as expressionless as a mask. Alfons Maria Jakob found similar symptoms of dementia in five men and women under the age of 50 in the 1920s. By the 1950s, white coats regarded classic Creutzfeldt-Jakob disease (CJD) as a rare syndrome that erupted sporadically in 60- or 70-year-olds. They thought it killed only one in a million people.

Meanwhile scientists were looking closely at scrapie. In the 1930s, French researchers injected a mixture of spinal cord tissue from a scrapie-infected adult sheep into a newborn lamb. It died 16 months later of madness, grinding its teeth. Farmers and scientists also nailed down the route of the infection: It was definitely cannibalistic. Sheep picked up the disease by eating placenta or placental fluids, a routine behavior in mammals.

In 1937 veterinarians also learned another important lesson about the transmissibility of TSEs. They took brain tissue from eight yearling sheep to make a vaccine to fight louping ill virus, a tick-borne invader then driving Scottish sheep crazy. The mothers of the vaccine fodder later developed scrapie, which meant that the yearlings were carrying the agent. One batch of the contaminated vaccine exposed 18,000 sheep to the scrapie agent, and nearly 1,200 came down with the disease.

This vaccine disaster put up some big red flags for British scrapie researchers. Not only was scrapie resistant to formalin, a pathogen-killing solution in vaccines, but the "infective scrapie agent could lie hidden in the tissues of clinically normal animals." So material from an infected few, diligently concentrated for industrial delivery to the many, could cause an epidemic.

By the 1950s, scrapie was looking more and more like a creature in a B movie that couldn't be killed. It remained infectious after being baked at 400 degrees Celsius (750 degrees Fahrenheit). Ultraviolet light didn't inactivate it. The agent could survive baths in chloroform and phenol. Even a brain removed from a dead sheep two years earlier still packed an infective wallop. The curious organism also appeared to contaminate the soil for years. Icelandic virologists noted that sheep free of *rida,* or scrapie, soon got sick grazing on fields previously occupied by sheep with the condition.

The next brain eater appeared among cannibals in New Guinea. The Fore, a Stone Age people, ate their dead in order to always have part of their dear departed inside them for the rest of their lives. Mothers and children got the choicest bits: brains partially cooked over a fire. Sometimes women even dug up the dead for a late-night protein snack. The first cases had appeared at the turn of the century, but by the 1950s hundreds of Fore people were starting to go mad. The Fore called the disease kuru, which means "to tremble with fear or cold." The Fore assumed that witchcraft spread the dreaded illness and bloodily dismembered anyone suspected of sorcery.

The keenly observant Fore wrote their own case definition for classic CJD by dividing the dementia into three stages. First came "walk-about-yet," when the limbs and torso of the sick twitched every which way. The eyes of the laughing or crying kuru victim would then cross. Next came the bedridden stage, or "sit-down finish." In the final phase, folk claimed by kuru lay as immobile as felled trees, unable to speak, staring vacantly at the world. The Fore called that part "sleep finish."

The Fore's brain-eating epidemic quickly became a medical curiosity. In the late 1950s, an Estonian physician, Vincent Zigas, started to study the phenomenon along with an ambitious American pediatrician, D. Carleton Gajdusek. The puzzled doctors noted that kuru afflicted as many as 5 to 10 percent of the women and children. Vigas and Gajdusek initially thought the disease was infectious, but no one with kuru ever had a fever or inflammation. Brain autopsies just showed gray matter full of holes, like those in a Parkinson's or Alzheimer's patient. "It is an astonishing illness," wrote Gajdusek. "To see whole groups of well nourished, healthy young adults dancing about, with athetoid tremors which look far more hysterical than organic, is a real sight." After examining some kuru brain samples, a European pathologist said it looked a lot like CJD.

Around the same time, Bill Hadlow, an American veterinary pathologist, started to study scrapie by slicing and dicing diseased sheep brains in England. The sheep trade had introduced the brain waster to the United States in 1947, and U.S. authorities wanted to learn more about it. After spending a year peering at spongy sheep brains, Hadlow concluded in 1959 that scrapie was something special in the world. Its long incubation period and infectivity simply astonished him. After he visited an exhibit about kuru in London, the disease's similarities to scrapie struck him like a thunderbolt, and he wrote a famous letter to *The Lancet* saying so. He recommended that bits of diseased cannibal brains be injected into monkeys to investigate infectivity. No matter how weird scrapie might be, it's "unlikely to be unique in the province of animal pathology," he wrote.

It took nearly six years and many dead primates, but by 1966 Gajdusek had transmitted kuru to chimpanzees. The animals trembled, shook, and died like the Fore. On the autopsy table their brains bore all the telltale signs of a TSE. Gajdusek then transmitted CJD from a human to a chimp with the same grim results. It was now pretty clear that cannibalism was recycling the disease agent among the Fore and that CJD didn't just appear sporadically.

While Gajdusek worked to amend the tribe's dietary habits, industrial farming continued to turn up more TSEs. In the 1960s, mink farmers in the United States found themselves dealing with fatal outbreaks of complete madness among their furry charges. The confused mink fouled their cages, bit off their own tails, dragged their hindquarters, and often died with their teeth firmly clasped to the bars of their cages. Farmers were feeding the carnivores slaughterhouse waste: dead cows, fish, cereal, or whatever was cheap and handy. The outbreaks in the 1960s were all traced back to "downer" cattle, animals too sick to walk. Scientists called the new invader transmissible mink encephalopathy (TME).

Shortly afterward the protein pantheon expanded again when another brain waster appeared in captive elk and deer at a wildlife research facility in Fort Collins, Colorado. Call this one a science-acquired infection. Bred for nutritional studies, a bunch of captive mule deer abruptly lost weight and showed signs of compulsive thirst in 1967. The animals slobbered and drooled and finally adopted the listless pose of mammals invaded by bad protein. Scientists can't say exactly what the animals were fed, but contaminated feed has never been ruled out. It's also possible that free-ranging deer had dined on meat and bonemeal powder in rations near a dairy farm and brought along the disease. Scrapie-infected sheep may also have played a role, but the evidence remains weak and inconclusive.

In any case, chronic wasting disease (CWD) exploded. For the next decade 90 percent of the deer kept at the facility came down with it. Researchers tried to rid the location of the infection by removing soil and drenching pastures and equipment with chlorine. But nothing worked; CWD just reinfected each new batch of science recruits. The agent also broke out of the facility and galloped through herds of nearby wild elk and deer. Their prion-rich urine and placentas infected more and more cervids throughout Colorado and Wyoming. A decade later, the wildlife pathologist Elizabeth Williams looked at brain slices from infected deer and noticed a profusion of distinct spongy holes. "Not many things cause that," Williams later told the reporter Philip Yam.

Although industrial animal feeding opened more pathways for the prion, medical progress opened some gates of its own. In the 1960s, researchers figured out how to extract growth hormones from pituitary glands by mining cadavers. For each production run, manufacturers combined pituitary glands pulled from the brains of 10,000 different dead people. By 1985 three recipients of the hormone had died of Creutzfeldt-Jakob disease. Before a synthetic hormone arrived on the market, more than 100 recipients went crazy. Contaminated brain grafts also spread more CJD around. So, too, did dirty surgical instruments.

Meanwhile the descendants of Liebig—those committed to better living through chemistry—kept promoting cannibalism, Liebig's "scientific feeding of animals." They reasoned that meat and bonemeal powder not only contained essential amino acids but also helped livestock grow faster. A herd of studies particularly pointed to how the MBM diet helped high-yielding dairy cows produce even more milk. At this point technology married political economy and gave mad cow disease a big-budget opportunity.

Between 1974 and 1983, generous government subsidies for the dairy industry encouraged British farmers to increase production. To get the job done, farmers diligently added more MBM to dairy cattle feed. The MBM diet for cattle and other ruminants rose from 26,000 tonnes (29,000 tons) to nearly 40,000 tonnes (44,000 tons) per year in one decade. By the time BSE emerged, 7 percent of the diet of most calves or dairy cows consisted of meat and bonemeal powder made from members of their own species. (In the United States, the percentage rose to 13 percent.) Mammals had never partaken in cannibalism on that scale before. In *Deadly Feasts,* the journalist Richard Rhodes astutely concludes, "An increased percentage of meat and bonemeal fed to cattle may in fact have triggered the epidemic." It also explains why 70 percent of all BSE cases occurred in dairy herds: They were the ones fed the richest protein diets.

A bit of infected tissue the size of a peppercorn probably started the English madness. (Invaders never start big.) A few isolated cases popped up in the 1970s but escaped detection. Farmers unwittingly dispatched these enfeebled beasts to the local renderers, who efficiently recycled the agent as more MBM. Each contaminated cow had a TSE payload big enough to infect another 10 cows. And that's all it took to stop a multi-billion-dollar trade, rewrite the menu in Europe, and contaminate the world's food chain.

The government might also have contributed to the mess with its lengthy insecticide campaign against warble flies. It used phosmet, a well-known neurotoxin. After MBM, phosmet remains the second-highest risk factor for BSE. Out of 169 British BSE cases reported in 1986, more than two-thirds had been treated with phosmet and all had eaten meat and bonemeal powder. The fact that more than 45,000 cows came down with BSE after two bans on feeding all animal protein to cattle (the first in 1988 and a stronger measure in 1996) indicates that a variety of environmental factors may influence the behavior of bad proteins. Mark Purdey, an English dairy farmer and lay scientist, has long argued that mainstream scientists just haven't explored environmental factors well enough. Most beef farmers readily second that assessment.

As BSE quietly infiltrated British beef herds and British beef eaters, Richard Marsh, a University of Wisconsin veterinarian, found another TSE in the United States. After a severe outbreak of TME that killed more than half of 7,000 crazed mink in 1985, Marsh diligently investigated their food source: 17 downer cows from a nearby dairy farm. He mashed up some infected mink brain and inoculated two healthy bull calves. The cows became listless and weak but didn't go mad. They eventually collapsed and died with very small holes in their brains. Then he fed mink some of the infected cattle brain and they, too, died. Marsh concluded that "a scrapie-like agent" was present in American cattle. North America, in other words, had its own BSE strain. The beef industry belittled Marsh's findings while the British uproar over mad cow quickly overshadowed them.

By the early 1980s, British farmers and veterinarians started to see hyperexcitable cows with uncoordinated gaits. The animals rapidly lost weight, got extremely violent, and then died as listless as women with kuru, sheep with scrapie, mink with TME, deer with chronic wasting disease, or humans with CJD. But nobody firmly diagnosed a spongiform encephalopathy until 1985. The first sample came from a cow that went berserk and trampled two sheep to death. Ten months passed before veterinarians acknowledged the existence of bovine spongiform encephalopathy (BSE); another seven months went by before politicians were advised, and another nine months elapsed before the health department got involved.

In fact the British response to mad cow disease set a remarkable template for how one government after another, from France to Canada, would respond to the invader. Commerce invariably prevailed, veterinarians made public health decisions, and the real experts on TSEs were systematically ignored. In the absence of good policy, politicians simply ate a lot of beef without understanding the science of brain-eating diseases or their long-term public health implications. Richard Lacey, a British microbiologist and food safety specialist, repeatedly characterized the mess as fraudulent and shameful. During the crisis Lacey frequently accused England's Ministry of Agriculture, Fisheries and Food (MAFF) of behaving like Orwell's "Ministry of Truth."

The initial ban on feeding any cattle bits to cattle didn't take effect until 1988, and it was so sloppily enforced that another eight years elapsed before the government closed all the loopholes. (In 1996 the British even made it illegal to keep meat and bonemeal powder made from ruminants in the same room with other livestock feed.) Meanwhile farmers simply used up their old feeds while renderers dumped their excess MBM on the international market, exporting it to Europe, Russia, Indonesia, Thailand, and India. No one in the British government thought of labeling the material as contaminated or poisoned. As one civil servant later admitted to the United Kingdom's BSE Inquiry, it was up to importing countries "to decide whether to import or under what circumstances." The government's libertarian decision pretty much ensured that the world's cattle business would be writing memos about BSE for the next century.

The British beef-eating public didn't really panic until cats started to die. People were still gobbling down meat pies even when the BSE caseload reached as many as 300 mad cows a week. But Mad Max changed that. The Siamese cat started to twitch and lick its skin bare in the spring of 1990. Then it dragged its hindquarters about like a mad mink. Hundreds of other felines and zoo animals fed pet chow made from BSE-infected cows later did the mad English dance. When the nation's chief veterinary officer announced that Max's death was "no cause for alarm at all," a third of the population used their own judgment. They observed that BSE had just done what the government said it would never do—cross the species barrier—and they stopped eating beef.

The British government then added to its mad cow follies. Civil servants with no expertise on TSEs assumed that the outbreak would decline, falsely blamed the whole thing on scrapie, and argued that cows were "dead-end hosts" that couldn't transfer the disease to humans. In an attempt to cool public angst, John Gummer, the minister of agriculture, tried to feed his reluctant daughter a hamburger in front of TV cameras. She politely declined to be a guinea pig.

The meat industry and its servile government friends then tried another tack. First beefcrats in England changed reporting procedures to distort the number of BSE cases and to minimize the scale of the invasion. The European Union's director general for agriculture, Guy Legras, penned a blunt memo in 1990 that read, "BSE: Stop any meeting." Other continental veterinarians recommended taking "a cold attitude towards BSE so as not to provoke unfavorable market reactions.... This BSE affair must be minimized through disinformation."

To accomplish that goal, the British government systemically ignored scientists with any knowledge of brain eaters. Hugh Fraser, who understood how TSEs could cross the species barrier and appear in different guises, was told to shut up. Alan Dickinson at Edinburgh University, perhaps the world's foremost authority on scrapie, got so disgusted with government shenanigans that he took early retirement. He later accused authorities of turning BSE into an epidemic by ignoring real scientists and by systematically weakening "the autonomy of UK science during the 1970s and 1980s." Authorities also dismissed Harash Narang, another microbiologist and TSE expert. He later wrote, "The handling of the BSE crisis was influenced by the preferences and prejudices of civil servants whose main concern was to protect the well-being of the farming industry."

In 1993 two dairy farmers died of CJD on BSE-infected farms. Next came a 15-year-old girl, Victoria Rimmer. She lost weight and slipped into a coma. After an autopsy confirmed classic CJD, a TSE supposed to affect only old people, government officials told her mother to keep quiet. "Think about the economy and think about the Common Market." Twenty-three-year-old Clare Tomkins fell ill in 1996. "The most harrowing thing was sometimes in bed at night," recalled her father during the BSE Inquiry. "She howled like a sick injured animal. She looked at you as though you were the devil incarnate. She started to hallucinate."

In 1996 the British government finally announced that 10 people had gone mad by eating contaminated beef. They called the world's newest brain eater "variant" CJD. Unlike classic CJD, the variant killed teenagers and young adults and ate up their brains rapidly. That announcement eventually toppled a government, led to the creation of a Food Standards Agency, and forced a total ban on feeding mammal protein to mammals. By now the BSE scandal had inflicted more than $10 billion in losses on the British economy.

After the stunning British revelation, the European Parliament set up an independent committee to look into the spread of BSE. Its 1997 findings didn't mince words. The report accused the British government of suspending veterinary inspections during critical periods, misleading trade partners, blatantly withholding information, and dumping contaminated meat and bonemeal feed on international markets with no warning. "It is clear that the policy of disinformation was not confined to misleading public opinion but played a full part in relations between Community institutions," wrote an indignant rapporteur, Manuel Medina Ortega. "It is surprising that the responsibility of producing a defective product and thereby causing a catastrophe on the beef market, has not been laid more firmly at the feet of the animal feed producers in the UK," he added.

By 2000 the basic lessons of BSE were as elementary as a glass of milk. Michael O'Brien, a British physician, spelled them out in the *International Journal of Epidemiology*. He advised that agriculture abandon the path drawn by Liebig and let animals eat their natural foods instead.

He also recommended that governments keep crazed animals out of the human food chain and, for that matter, trade promoters out of public health or agricultural agencies. Good disease monitoring and timely research were also good tools. Last but not least, it didn't hurt to "tell the truth." These lessons, however, proved too hard for nations dependent on beef dollars to swallow.

In fact, every country hit by BSE has had trouble telling the truth. This unease with real science has, in turn, amplified the domain of the brain eaters. Both European and North American politicians repeated the British experience with such obliging deference to trade that it's tempting to conclude that they all read the same manual on how to propagate invaders.

Consider France, which imported lots of BSE trouble from nearby England. It began its mad cow tango in 2000 with meat recalls, bans on cattle protein in feed, and the mandatory testing of 20,000 animals a week. The families of two dead meat eaters with variant CJD promptly sued the government, while farmers took to the streets to demand compensation for killed cattle. A year later a report by the French senate concluded that the agriculture ministry had minimized the threat of mad cow and "constantly sought to prevent or delay the introduction of precautionary measures."

Germany, home for hundreds of varieties of beef-filled sausages, then got clobbered. Its government claimed that mad cow could never happen in the Fatherland. When a suspicious beef producer did his own testing and found lots of cows with holes in their brains, all hell broke loose. The nation's "head-in-the-sand policy" quickly forced the resignation of both the health and agriculture ministers. To restore confidence to the marketplace, the government later committed a fifth of all farms to green practices that eschewed agricultural cannibalism.

Japan was the next naked emperor to feel the BSE chill. In 2001 it was still feeding animal bits and cattle blood to its bovines and importing MBM feed from Europe. So BSE, whether local or imported, had ample means of recycling itself through Japanese herds. Although European scientists had warned the country's die-hard beef eaters that Japan was a "high-risk" country, the government pretended the risk was "minimal."

The first case of BSE appeared in a five-year-old Holstein in the Chiba prefecture, a regional feedlot for 1.5 million animals, in late August. While disbelieving officials rechecked the test, the carcass of the infected animal went into a 130-tonne (145-ton) batch of meat and bonemeal powder. Before flustered authorities could cull 5,000 animals that had dined on the contaminated feed, another two BSE cases rained on Japan's beef parade. Beef sales dropped by 60 percent. Just three cases of BSE had cost the economy nearly $3 billion.

In response, the Japanese health minister, Chikara Sakaguchi, and the agriculture minister, Tsutomu Takebe, grabbed their chopsticks and publicly ate beef. When the government then offered to buy up thousands of animals, opportunists rushed in. One company, Snow Brand, imported 12 tonnes (13 tons) of Australian beef and repacked it with worthless Japanese beef in order to qualify for the big government compensation package. That scandal eventually led to the resignation of the assistant farm minister.

In 2002 an investigation accused Japan's agriculture bureaucracy of "serious maladministration" and concluded that it had "always considered the immediate interests of producers in its policy judgments." That same year the meat inspector who discovered the nation's fourth mad cow committed hara-kiri. Hundreds of small butchers went out of business. To restore confidence in the beef market, the country incinerated $160 million worth of beef and introduced the world's most comprehensive testing system. It soon detected BSE in cattle as young as 21 months, another development that wasn't supposed to happen. Only cows at least 30 months old could carry the disease, civil servants had concluded.

North America's integrated beef market soon repeated the global

BSE follies. Canada and the United States profited mightily during Europe's BSE crisis by touting their apparently "BSE-free" status. The two major beef exporters demanded seven-year-long bans on imports from countries with one mad cow and then took over their beef markets. North America's BSE-free status, however, was a bogus claim defended by ignorant civil servants, minimal testing, and inadequately enforced bans on meat and bonemeal feed. Critics accurately called it a "don't test, don't find" system.

Canada, a colonial producer of beef for American slaughterhouses, set the tone. Its agricultural officials dismissed BSE as a "European disease," encouraged the export of nearly a million live animals a year, and advertised as "rigorous" a testing regime that examined only 800 animals a year.

But a secret 2000 report by two toxicologists for Health Canada warned the government that the northern beef factory probably had lots of brain eaters. The country had imported contaminated MBM as well as slaughterhouse waste from France, a country with a big BSE problem. It also noted that Canada's rendering industry had put everything but the kitchen sink into ruminant feed prior to the nation's 1997 "voluntary" ban on cattle protein (which wasn't fully enforced until 2003). For most of the 1990s, dairy cattle got to eat bits of cow spines and heads, roadkill, euthanized dogs and cats, the offal of pigs and sheep, supermarket wastes, diseased animals with TB, and even farmed elk and deer with chronic wasting disease. The cooked-up protein went to 500 feed mills or 10,000 individual feeding stations on farms. "The possibility that a BSE inoculation event occurred in Canada … must be considered," the toxicologists wrote. Beef promoters in government, however, prevailed, and the report was never publicly released.

When three mad cows popped up in Alberta in 2003 and 2004, Canadians got a taste of the British experience. After the first cow went mad, Prime Minister Jean Chrétien sat down to a beef dinner and the premier of Alberta, Ralph Klein, advised "self-respecting" farmers to shoot, shovel, and shut up. After the second cow appeared, the Canadian Food Inspection Agency, a regulator with a dual mandate to promote trade and protect food, chanted, "The system was working." Although the government refused to consult with major TSE experts, it still called its poor testing program "science based." The U.S. government offered the same propaganda on its first two cases, and its agriculture secretary, Mike Johanns, also sat down to "enjoy a good steak." Meanwhile North America's ordinary cattle producers lost $8 billion while multinational slaughterhouses raked in record profits.

To date the United States Department of Agriculture (USDA) has a record on BSE as appalling as that of its Canadian counterparts. In 2002 the General Accounting Office noted that the U.S. ban on cattle canni-balism that had gone into effect six years earlier was deeply "flawed." In 2003 a report by a panel of European BSE experts lambasted the Americans for inadequate testing, for refusing to adopt rapid European tests, for not doing enough to protect consumers, and for not rooting out the diseased cows. It said telling people that beef is safe is no substi-tute for policies that ensure it is safe.

A 2004 investigative report by *The Guardian,* a London newspaper, found more of the same. It concluded that U.S. policies generally favored industry at the expense of consumer safety, that testing for BSE was rare and haphazard, that discussion of the disease was discouraged, and that government agencies didn't follow the rules. As one American vet put it, "Among ourselves we think our inspection system is the lowest in the world."

Remarkably, the inspector general of the USDA later came to the same conclusion. In 2005, she found that the agency didn't test the riskiest animals and had in fact misplaced or lost 500 cattle brains waiting for testing. Confusion also reigned among inspectors, because of sloppy training and incredibly poor record keeping. In 2006 the inspector general again noted that the USDA had gone out of its way to resist proper testing on another suspected case of BSE: "As a result, should serious animal disease be detected in the United States, USDA's ability to quickly determine and trace the source of infections to prevent the spread of disease could be impaired."

The North American strategy for dealing with BSE so far has mostly been shoot, shovel, and shut up. The ongoing scam does not surprise Margaret Haydon. Health Canada fired the veterinarian drug evaluator and two other prominent scientists for their outspoken support of the real science on BSE. She recalls that one brave American vet diagnosed two brain-eating cases in the early 1990s but to no avail. "The federal government screamed for the tissue samples and [the result] never came out. They've got it and we know it. It's another case of corruption."

Nevertheless Canadian and American civil servants as well as the upper echelons of the cattle industry pretend our meat is safe. But given the durability and flexibility of TSEs, Haydon and many other experts aren't so confident. "On what basis? The prions are there if there is an infected carcass. How can you say it's not in the meat?" No one is testing steaks and roasts for BSE, but many experts suspect that if grocery stories did, they would surely find bad prions.

To ensure that mad cows remain out of the headlines in North America, the U.S. government plans to reduce its passive testing program to 40,000 tests annually, or 110 a day. Marc Savey, research director for the French food safety agency, has called that small number "ridiculous." A 2004 French study published in *Veterinary Research* demonstrated that the inadequate surveillance program adopted by North American governments would allow a BSE epidemic to go unrecognized and underreported by 80 percent. The French scientists know because poor testing is what the French government did until 2000. North American consumers remain the only beef eaters on the planet who have not demanded a shake-up of their agricultural bureaucracies or their beef monopolies.

Although no one wants to talk about the public health implications of this mess, they remain either disturbingly significant or significantly unknown. In the United Kingdom, beef eaters consumed approximately 54 million doses of BSE-infected meat during the epidemic, and initial predictions suggested tens of thousands might go mad. Now most scientists gingerly suggest that no more than 500 Europeans, including Britons, will die from the human strain of the disease. There are good reasons for the welcome downsizing. One explanation may involve the amount of infectious material consumed. In this respect, BSE may behave like scrapie: Studies in mice show that it takes repeated doses of a little infectious material over long periods to establish a brain eater. A single larger dose doesn't induce disease. In addition to sharing a genetic susceptibility, the unfortunate human victims of the BSE epidemic may have simply eaten more contaminated hamburgers than the normal population.

Yale professor Laura Manuelidis has another explanation for BSE's muted killing powers. In 2005 she found that mouse brain cells infected with two strains of scrapie couldn't be infected by variant CJD. In other words, 300 years of exposure to scrapie in England may have created a barrier to mad cow disease. "Maybe we've all been infected already with scrapie and that's why we have not seen more cases of vCJD," muses Stephen Dealler, a microbiologist. Perhaps scrapie has saved Europe from a massive outbreak of neurodegenerative disease. North Americans, less frequent mutton diners, may not be so lucky.

While BSE gets most of the attention, other TSEs continue to wade through deer and elk populations in North America. Chronic wasting disease has now occupied 11 states and two Canadian provinces. It's even on the doorstep of New York City. The factory farming of cervids for the aphrodisiac trade ("velvet Viagra") has moved the agent around with the help of government types committed to free agricultural trade. Outbreaks of CWD have occurred on 43 game farms or deer factories in Saskatchewan and Alberta, costing Canadian taxpayers tens of millions of dollars. As in chicken factories, the density of these operations predictably invited mass mortalities ranging between 50 and 90 percent.

Escaped animals, or what scientists call "spillover" from factory farms (and deer do like to jump fences), have now spawned a major epidemic among free-ranging deer and elk in the Canadian west. That outbreak, combined with widespread spillover events in the United States, imperils North America's $10-billion wildlife viewing business. The prospect of seeing, let alone photographing, emaciated mammals that drool uncontrollably with bowed heads could level tourism in American and Canadian parks the way foot-and-mouth killed visits to rural England.

A 2004 Canadian report by a seven-member team of TSE experts laid out a familiar story for CWD. Trade in infected deer or "trans-

CHRONIC WASTING DISEASE IN NORTH AMERICA

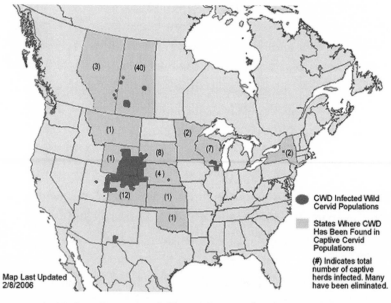

Trade in factory-raised deer has driven chronic wasting disease across North America in trucks and feed sacks.

Source: Chronic Wasting Disease Alliance. Reprinted by permission.

location of captive and free-ranging animals has resulted in CWD range expansions and once established, the disease may be maintained through environmental contamination for an unknown period of time." The report also found that there were no natural barriers to stop CWD in the wild. So this brain eater may soon spread to cougars, bison, bighorn sheep, and moose. Nobody knows how many strains of CWD there might be. "This is a catastrophe in the making, and there have been very few resources allocated to deal with it," declared the University of Saskatchewan biologist François Messier in 2004. Since then various Canadian and American governments have started more eradication campaigns for wild deer to contain the madness. Somehow trade in biological invaders always ends up with mass murder of one kind or another.

Meanwhile the world of prions continues to be a world of wonder. A sheep in Vermont recently acquired BSE from contaminated animal feed. Some men who ate squirrel brains developed holes in their brains, and young hunters who ate venison also died of brain-wasting diseases. Mysterious clusters of undefined CJD in young people have been reported throughout the United States. Idaho reported nine cases of classic CJD in 2005; according to medical statistics, the state should yield only one case a year. At annual conferences the families of those killed by CJD in the United States say prayers of remembrance and lobby to make it a reportable disease.

The number of Americans recorded as dying from Alzheimer's jumped from 653 in 1979 to 60,000 in 2004—a trajectory that closely follows the use of meat and bonemeal powder in animal feeds. But many of these patients are actually dying of some form of CJD. In 1989 Laura Manuelidis found that up to 13 percent of patients diagnosed with Alzheimer's by competent neurologists were found to have CJD when cut open on autopsy table. A University of Pittsburgh study the same year found that 6 percent of dementia patients had CJD. Another study on spongiform changes in the brain found that samples from 50 out of

66 patients with Alzheimer's were virtually indistinguishable from samples showing CJD. Given that fewer than 5 percent of the dead are ever autopsied anymore (and CJD can be confirmed only by autopsy), a stealth infection of brain eaters could be eating through North Americans. Out of 4 million diagnosed cases of Alzheimer's disease in North America, 120,000 may actually have CJD, caused by eating beef infected with a strain of BSE. "If we don't do autopsies and we don't look at people's brains ... we have no idea about the general prevalence of these kinds of infections and [whether] it is changing," argues Manuelidis.

In England the BSE story goes on and on. Dirty feed bins continue to pass on the bad protein, and sloppy abattoir testing permits contaminated meat to travel from farm gate to dinner plate. The persistent BSE prion has even found a way to pass naturally into sheep. Scientists have also discovered that more than 3,000 Britons could be incubating variant CJD in their tonsils and tongues. Dentists now take greater precautions when examining their patients' mouths.

Brain eaters are no longer found only in the brain or spinal cord tissue. Just about any organ, from liver to kidneys, can support enormous loads of prions. They also tend to concentrate in any area that has suffered trauma. In the United States, the prized hindquarters of infected deer and elk have been found to be so riddled with prions that experts are at a loss for words. "It's frightening that there should be so many prions in the muscle meat of deer," reflects Stanley Prusiner. Some scientists now fear that if CWD ever jumped the species barrier it could be transmitted by something as innocuous as a kiss.

Around the world 80 countries are heedlessly incubating BSE in their cattle. Most, such as Indonesia or Thailand, don't have the money to test their beef and so will recycle the agent ad infinitum. According to the International Feed Industry Federation, production of animal feed from both animal and plant sources will treble by 2050 in order to supply *Homo economicus* with his favorite meat dishes. That inevitably means more concentration of proteins, cross-contamination, and inadequate enforcement of feed bans. In other words, global trade in contaminated feed can't help but launch more TSEs.

It has already launched tonnes of other invaders from farm to fork. Just 10 countries now account for 60 percent of the globe's animal feed production, and they import and mix the raw materials (grains, legumes, and rendered animal parts) from all over the world. Recent studies by the FAO prompted by the mad cow debacle have found bacteria, antibiotic growth promoters, and fungi and their poisonous mycotoxins in animal feed. One investigator summed up the the whole industrial mess as "contamination from diverse sources." The bacterial hitchhikers included a host of well-known food-borne diseases: *E. coli,* 2,000 different kinds of salmonella, *Listeria monocytogenes,* and campylobacter. A 2002 study published by the journal *Clinical Infectious Diseases* spelled out the obvious: The microbial contamination of animal feed remains an open door for "the entry of pathogens into the human food supply."

Analyzing the content of animal feeds is both difficult and expensive, but whenever reluctant regulators have taken the time, they've discovered a lot of unlisted invaders. In 1998 an American study found that 30 percent of cattle feed in the United States contained *E. coli* germs. Between 1996 and 2000, the U.S. Animal Protein Producers Industry found salmonella in one-quarter of its products. European studies have also discovered dangerous levels of dioxins, radioactive particles, antibiotic residues, and pesticides. Given this kind of leakage into the human food chain, it would be surprising if TSEs weren't making their way to the human plate on a global scale every day. An understated 2004 report on the status of animal protein in livestock feed by the FAO briefly concluded that "there is a growing need for transparency in the animal food chain and continuing vigilance."

In 1997, at the height of England's mad cow frenzy, an American physician, Robert Klitzman, made a trip to the highlands of New Guinea. Nearly 40 years before, Klitzman had worked with Carleton Gajdusek to solve the kuru puzzle. The Americans had done a good job. After the Fore stopped eating brains in the 1960s, kuru had almost disappeared, except for occasional transmissions from mothers to children, a legacy that BSE may yet repeat in England.

While in the Fore region, Klitzman was surprised to meet a man who seemed shaky and tremulous. Klitzman thought it looked like kuru and asked how old the man was. The man didn't know. But he reckoned that he was seven or eight years old when a road had been built in the area in 1957. That pegged him as 47 or 48 years old. He said that his mother died of kuru and that as a child he had shared human brains with her at a funeral feast.

That revelation amazed Klitzman. It meant that a meal consumed four decades earlier had come back to devour the man's brain and end his life. He called it the longest incubation period for any pathological protein in human beings, or in any other species, for that matter.

The physician then paused to consider the implications: Hamburgers or meat pies eaten during England's mad cow crisis could haunt people for another three or four decades. "Here in this village with its dirt, bare huts, and unwashed feet, a scientific record was being set."

Given the world's appetite for meat and the deceptive practices of global traders, it won't be the last chart-buster set by a TSE. The invasion of the brain eaters has just begun.

FIVE

Rusts, Blights, and the Invaded Larder

"Virtually every one of us faces the consequences
of our ignorance of agriculture three times a day."
—Richard Manning

C offee, like every major crop favored by humans, tells a fabulous
and aromatic story about biological invasions and the appalling
state of modern agriculture. It begins with the famous bean that evolved
in the highlands of Ethiopia. There long-legged tribesmen chewed on
the caffeine-rich berries for a dependable energy fix while hiking over
mountainous terrain. The plant didn't make its way to Yemen until
1000 AD, along with an invasion of Ethiopians. Mullahs soon cultivated
the shrub, dried the berries, and brewed the first cup of arabica to
sustain long days of praying. Coffee was such a stimulating drink that
Muhammad once boasted that he "could unhorse forty men and possess
forty women" after a jolt of joe.

The spiritual bean soon spread to Persia and Turkey. After much
scheming, Europeans liberated the plant from the Middle East in the
16th century and the bean went global. The Dutch, among the world's
most commercial-minded flora and fauna arrangers, transplanted the

crop throughout the East Indies and Ceylon (now Sri Lanka). Before the plant could produce a profit, the Dutch lost their colony to the British Empire, an emerging political organism that proved pathogenic to African and Asian societies.

Modern agriculture then played the first of its many discordant melodies. To profit from the growing popularity of coffeehouses in Europe, the British government encouraged the expansion of one fine-tasting species, *Coffea arabica,* in the Ceylonese highlands. Free of African predators, the trees flourished in the tropical environment, and by 1870 Ceylon had become the world's biggest coffee kingdom.

It was a biological invasion full of cultural and global ironies. To refill coffee cups in Europe, where German girls described the drink as "lovelier than a thousand kisses," an Ethiopian tree shrub displaced more than 160,000 hectares (400,000 acres) of virgin forest. When Sinhalese workers proved insufficient for the task of cutting native vegetation to make room for the imperial tree, the coffee magnates imported thousands of Tamils from southern India. (Biological invasions have a way of sowing the seeds of future political and economic explosions.) Soon Ceylon had more coffee than local oxen could carry to port, and British railway engineers set to work.

No one questioned the moneymaking boom except the superintendent of the colony's Royal Botanic Gardens, prickly and sullen Dr. H.H.K. Thwaites. "Any country that gambles its future against the quick wealth that comes from a single crop is inviting bankruptcy," warned the spoilsport. Commerce, however, triumphed over Thwaites's prudence, and Ceylon became one big vulnerable coffee plantation.

Pathogenic bankruptcy did arrive in the form of *Hemileia vastatrix,* the coffee rust fungus. Fungi, which have more than 1 million species, are the earth's number one garbage recyclers and decomposers: They decay dead things (the planet would be unlivable without them) and return various essential nutrients to the soil. With their weblike structures (called hyphae), fungi also extend the reach of plant roots to collect more water and nutrients. Only 10 percent of the world's fungi have been identified, so scientists and customs inspectors have never had enough information to build solid defenses against fungal invaders.

Mobile fungi, much like rootless humans, can behave as shamelessly as zebra mussels. In fact, the majority of the world's most devastating crop diseases are fungal in origin. They have probably killed more people—by robbing them of dinner—than all recent plague celebrities combined, including AIDS, West Nile, SARS, and bloody Marburg.

Hemileia vastatrix probably originated in Africa, where the diversity of coffee plants and other trees kept fungi in check. But Ceylon's unshaded rows of a single species of coffee offered no barriers, just an unparalleled feast. Billions of yellow-orange fungus spores were carried about the island on the wind, insects, and Sinhalese workers. Upon landing on a leaf, a spore would germinate, suck the life out of the plant, and breed more spore-producing blisters called pustules. So a coffee monoculture turned Ceylon into a highly efficient fungus factory.

At first the planters ignored the disease and simply planted more trees. When the invader also stripped the replacements of energy, desperate planters called for science, inquiries, and action. In response the British government sent out Harry Marshall Ward, one of the fathers of the science of plant diseases and a genuine naturalist, to assess the situation. Ward sectioned leaves and counted the pustules while he marveled at Ceylon's sensual scenery. When the planters demanded chemicals, the scientist gave them the sobering truth. In his famous report on the "coffee leaf disease," he noted that specialized cultivation of one crop over a wide area amounted to one hell of a way to brew disease. By hastily stripping the hillsides of all native vegetation and tall trees, the planters had removed a critical disease filter.

Within a decade the rust buried the coffee industry. Exports fell from a high of 45 million kilograms (100 million pounds) in 1870 to just 2.3 million kilograms (5 million pounds) by 1889. The fungus went on to destroy most arabica plantations throughout Asia. After coffee rust conquered Ceylon, political implosions unsettled island life. Planters went bankrupt and when their key investor, the Oriental Bank, collapsed, thousands of naive coffee investors lost their savings as well. The Royal Navy transported thousands of destitute Tamils back to India in a forced migration dictated by a fungus. Sinhalese workers couldn't afford rice, let alone coffee. And in Europe coffee drinkers gave up their high-priced drink and switched to tea. The coffee trade moved on to *Hemileia*-free South America. When the fungus eventually showed up in 1970, the Brazilians burned an area 65 by 800 kilometers (40 by 500 miles) in an unsuccessful attempt to beat back the invader. A simple fungus can create a hell of a rumpus.

M odern agriculture has now done to much of the world what coffee did to Ceylon in the 19th century. As a consequence, almost every crop of any nutritional or economic significance is now under siege or has had a brush with death. Monoculture and limited genetic diversity combined with unprecedented trade in crops and seeds have given a veritable horde of fungi, viruses, bacteria, and plant-eating insects a competitive edge. When Tommy Thompson retired in 2004 as the U.S. secretary of health and human services, he famously remarked, "For the life of me, I cannot understand why the terrorists have not targeted our food supply because it is so easy to do." The simple answer is, they don't have to. Global trade and our cosmopolitan habit of eating stuff from other people's backyards have accidentally set off a chaotic string of biological explosions that no terrorist group could deliberately replicate.

In fact, crop invasions around the world are now almost as routine as suicide bombings in Iraq. In California, wine growers are battling an old bacterium carried by a newly introduced insect invader called the glassy sharpshooter. A hardy new strain of black stem rust (Ug99) is reducing wheat yields in Africa by 70 percent and is on its way to becoming a global grain bane. Meanwhile fusarium head blight continues to wilt crops in North America's breadbasket. Throughout Europe, peach and apricot growers are battling plum pox (a virus that appeared after World War I and that can lay waste to 80 percent of an orchard) by burning millions of trees. The bean dwarf mosaic virus could chew up the soybean in Argentina the same way *Hemileia* cooked the coffee bean in Ceylon. Sugarcane orange rust terrorized Australia's sugar production throughout the 1990s. Each year rice blast disease gobbles up enough of the Asian staple to feed 60 million people. Sorghum ergot, a toxic European pathogen, has traveled on the heels of the global hybrid seed trade to jeopardize the production of livestock feed throughout the New World. A severe form of mosaic disease has rotted out cassava through-out Africa, causing famine in Uganda. Geminiviruses, transmitted by whiteflies, are gobbling up tomatoes, melons, and cotton crops. The citrus canker, a bacterial invader, has put Florida's citrus industry in peril and prompted the destruction of 2 million trees. High plains disease, a virus that appeared in 1993, is now threatening to become a serial killer of wheat and corn crops. Meanwhile soybean rust, a Japanese native, has become a global villain and a threat to the multibillion-dollar U.S. soybean industry. A concerted fungal attack could make the last commercial strain of banana extinct, and even cacao, the source of chocolate, is facing a biological meltdown.

Plant pathologists, who are used to the natural phenomenon of diseases and pests raiding our crops, are becoming anxious. At a collo-quium on "new and emerging plant viruses" in 2003, they debated whether the microbial world is mounting a concerted attack on domes-ticated plants or whether the movement of plant material (trade in seeds alone is worth $40 billion a year) and flexible trade barriers have given invaders an unnatural boost. But everyone agrees that the global exchange of agricultural pathogens is real and growing. In fact, biologi-cal invaders alone now account for about 60 percent of U.S. crop losses every year, at a cost of $137 billion.

THE GLOBALIZATION OF SOYBEAN RUST

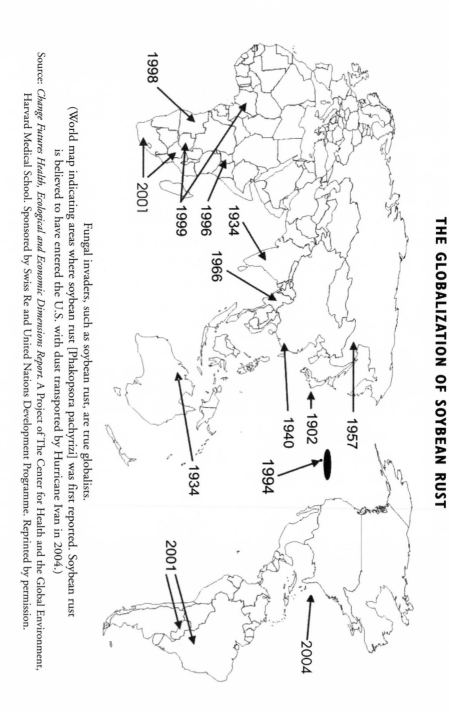

Fungal invaders, such as soybean rust, are true globalists.
(World map indicating areas where soybean rust [Phakopsora pachyrizi] was first reported. Soybean rust is believed to have entered the U.S. with dust transported by Hurricane Ivan in 2004.)

Source: *Change Futures Health, Ecological and Economic Dimensions Report*. A Project of The Center for Health and the Global Environment, Harvard Medical School. Sponsored by Swiss Re and United Nations Development Programme. Reprinted by permission.

In Australia, plant pathologists can't sleep at nights. Members of their depleted ranks are now wondering how they can ever protect 300 valuable commercial crops from as many as 3,000 potential exotic invaders when every month quarantine officials confiscate 3,000 seeds, plants, and fungi at the post office alone. Although they know good science and a vigilant infrastructure are necessary, governments and traders aren't keen on either, notes Lester Burgess, president of the Australian Plant Pathology Society. Quarantine "runs smack into conflict with the rules of the World Trade Organization and our competitors in the European Union and the United States, who see our quarantine policies as being against the spirit of free trade." It's a common complaint among plant doctors around the world. Most admit that they cannot successfully fight exotic invaders and apathetic policymakers at the same time.

Ironically, the huge vulnerabilities unmasked by biological invaders have created a flurry of interest in the expanding field of agroterrorism. The United States, for example, has 400 million hectares (1 billion acres) in crop and range land, which makes finding a new invader as easy as spotting a needle in a haystack. Reports to Congress and articles in science magazines now lament how "pest and disease outbreaks can quickly halt economically important exports" or undermine faith in government. Some scientists have even penned detailed accounts of how to poison North America's milk supply with bacterial toxins. Other alarming documents on invasive threats have concluded that "if a pest can enter the United States, over time, it will find a way here." The same goes for the globe.

Agriculture has always been a heroic epic about man outwitting local pests, disease, and the weather. But modern agriculture, built on the conceit that food can be grown and exported anywhere, provided we back up the enterprise with the technological equivalent of the U.S. Marines (think pesticides, fungicides, herbicides, fertilizers, and corporate seed trades), has turned this tale into a tragicomedy in which we openly invite global pests (and highly resistant ones at that) to play leading roles. What has always been a difficult task may well be on the verge of becoming a constellation of minor and major disasters that could all end as badly as coffee in Ceylon.

Putting dinner on the table has always been a tenuous and unstable enterprise and man's most invasive activity. Wheat and corn are to wild grasslands what smallpox was to aboriginal peoples: an indiscriminate eraser of diversity. Soybeans are to tropical forests what *Hemileia vastatrix* was to coffee; they are what scientists politely call "global change agents." Cultivated cropland now occupies an area of the globe about as big as South America. In other words, about a third of the world's green spaces have been rearranged to support our favorite cereals and vegetables. So agriculture is really the story of one man-managed invasion after another and nature's inevitable counterattacks. We are now witnessing what the Roman poet Horace recognized: that no matter how hard we try to drive nature out "with a pitchfork," it always returns.

The seeds of agricultural instability are ancient. Some 14,000 years ago, various human tribes abandoned the healthy life of hunting and gathering for the dusty work of plant breeding. It was not a rational exchange. The fossil record clearly shows that early farmers were a short, disease-ridden lot with none of the vigor of hunter-gatherers. But plant taming did allow the species to multiply and concentrate on quantity instead of quality. What we surrendered in good teeth and noble stature, we gained in sheer numbers. Agriculture simply allowed the species to expand and invade more parts of the planet.

No one really knows why we made this dirty gamble. But many weather experts now argue that a dramatic warming period, complete with melting glaciers, made agriculture possible on a large scale for the first time. The disappearance of big meaty mammals because of over-hunting might also have encouraged greener thumbs. As we ran out of mammoth steaks to put on the barbecue, we started to fool with plants (and to domesticate animals) in various cool uplands and mountainous regions from Papua New Guinea to northern China. In the process humans tamed 7,000 wild edible plants.

After selecting vegetation with the tastiest or largest seeds, kernels, fiber, and berries, we then spread the good news around with the efficiency of a spammer. None of the world's 103 major food crops (or their native diseases) have stayed in their original strongholds for long. The banana and the mango escaped Borneo and conquered the Pacific islands, China, and India. The orange abandoned the sunny hills of northeastern India to become a fruit of global commerce. In the Near East, a combination of wheat and barley (bread for man, barley for cattle) quickly displaced grasslands with irrigated fields and then conquered Europe, Africa, and India. In the last 400 years, the common amaranth of Mexico has been adopted by hill tribes in the Himalayas, no less, while Africa has gratefully received the cassava root and the potato from South America.

All of this intense plant taming and exchanging begat pyramid building, crowded cities, raging armies, and religions that saluted wheat or rice as godly gifts. In his sober history of agriculture, *Against the Grain*, Richard Manning argues, "We learned to grow food in dense, portable packages, so our societies could become dense and portable." With that density came all kinds of disease and climate vulnerabilities. Famine and pestilence have dogged agriculture's every tortured step. Whenever the rains fell in even measure, the species flourished; when they didn't, blights and rusts consumed our main cereals and we resorted to famine filler: our children.

To keep such uncomfortable protein choices at bay, farmers used to cultivate diversity and mimic nature. The practice not only provided crop insurance but meant peasants wouldn't have to eat their progeny. (Diversity, says the ecologist Paul Ehrlich, is man's "only defense against the unknown.") In India, for example, farmers at one time planted as many as 400,000 varieties of rice selected for their ability to withstand disease, drought, and other vagaries of life. But the industrialization of agriculture combined with the so-called green revolution traded off that insurance for higher yields from fewer than 30,000 varieties. When the choice was between growing a variety of crops that couldn't keep up with population growth and wide-scale crop homogenization that created a mirage of abundance, we made the obvious selection.

Joseph Stalin, the bloody Soviet gangster, once explained to Winston Churchill the totalitarian nature of this scientific reduction: "We have improved beyond measure the quality of our wheat. We used to sow all varieties, but now we only cultivate the Soviet prototype. Any other cultivation is prohibited nationwide."

This Soviet-like model has transcended all isms to become the global norm. As a consequence, monoculture, the planting of one crop variety over a large area, has pretty much won the day in agriculture. Corn used to come in 50,000 varieties suited for different locales and for resistance to particular pests, but agribusiness prefers just six types. Wheat, which once boasted 30,000 varieties, now comes in only three or four specialized varieties that must be engineered to withstand 135 kilograms (300 pounds) of pesticide poison per acre. The mighty field of 5,000 potato varieties has largely been reduced to the Burbank, an obese tuber suited to making french fries, which fatten both North American diners and fast-food profits. Most of the apple's 7,000 varieties are gone. Lettuce, once seen in a global array of sizes, shapes, and tastes, is now represented mostly by iceberg or romaine, which together dominate North America's fields. Although today's tomatoes couldn't be grown without the disease resistance they have acquired from wild species, the majority of their ancestors have been wiped out by land abuse. All in all, the Food and Agriculture Organization estimates that three-quarters of the genetic diversity in food crops has been washed away, dug up, or paved over, despite earnest attempts to preserve it in seed banks or protected areas.

The disappearance of these wild and traditional varieties—what scientists call land races—has been as disastrous as soil erosion. In many cases, diverse varieties representing an ancient biological inheritance have been destroyed by city building or clear-cutting. Farmers, who have been taught to behave like Internet tycoons rather than humble biological stewards, have also given up old domesticated varieties for cash crops with uniformly high yields. The plant geneticist J.R. Harlan sadly notes that some of the world's greatest centers of plant diversity have now become the most impoverished because of this practice. Harlan, who came from a family of plant collectors, never thought this man-induced genetic erosion was a terribly bright idea: "These resources stand between us and catastrophic starvation on a scale we cannot imagine. In a very real sense, the future of the human race rides on these materials." Garrison Wilkes, a retired professor of plant biology and maize expert, put it another way: "The extinction of local land races by the introduction of improved varieties is analogous to removing stones from the foundation to repair the roof."

The floor has caved in all over the place. It made one heck of a crash in the United States in 1970. That's when southern corn leaf blight, another invasive fungus, ruffled the tassels of the corn industry, which then produced 70 percent of all grain fed to livestock. The Illinois plant pathologist Albert Hooker later called it "the greatest loss in a single crop in a single country in a single year of any plant disease in recorded history." As any fungus might tell you, only a monoculture could achieve such great losses in so little time.

Throughout the 1960s, 80 percent of all hybrid corn planted in the United States contained the same genetic material: the "Texas male sterile cytoplasm." Plant breeders had used the gene out of laziness; it just made breeding and pollinating large numbers of plants less labor-intensive. But efficiency for guys in white coats tends to translate into something else for Mother Nature. Although plant breeders had reported as early as 1962 that the dominant hybrid seemed almost hypersensitive to the fungus *Helminthosporium maydis,* nobody paid much attention. And so the seeds of another man-made epidemic were quietly sown.

When a new race of *Helminthosporium* turned up in 1970, it dismayed farmers, upset the stock market, and even rattled President Richard Nixon. The epidemic began, as most fungal fiestas do, with unusually moist weather. Race T, or what scientists later called "an unforeseen mutation," started its campaign in Florida and then worked its way across the South and into the U.S. corn belt. Unlike previous varieties of the fungi, which nibbled modestly, Race T could reproduce in 51 hours after infecting a plant. It also invaded the whole plant and melted the stalk, husk, kernels, and cob. The uniformity of the crop was "like a tinder-dry prairie waiting for a spark to ignite it," wrote one pathologist. "Race T was the spark."

The blight destroyed 15 percent of America's corn crop: a billion dollars worth of protein, or enough corn to produce 7 billion half-kilo (one-pound) steaks. British farmers read the headlines and switched to barley for livestock feed, fearing unsustainable losses. Had the weather not cooled down, the fungus might have consumed more than half the U.S. crop. In Guatemala or Kenya, where the majority of people get half their calories from corn, such an event would have led to widespread famine for humans and livestock alike. Only America's unparalleled wealth and a lucky climate break saved it from disaster. The blight, however, prompted a major investigation by the National Academy of Sciences, which concluded that a "quirk in technology" had turned American corn into "identical twins." The NAS also noted that most other crops were "impressively uniform genetically and impressively vulnerable."

Although U.S. corn production now uses a greater variety of hybrids, it has become more and more uniform as farmers employ the same corporate seeds, fertilizers, and pesticides. Corn production, like most U.S. monocultures, behaves more and more like one big farm. "Although it may seem to scientists and government officials that we are now light years away in our knowledge and abilities from the 1970 Southern Corn Leaf Blight," writes the journalist Jack Doyle, "we are only in fact one unknown 'genetic window' away from a new agricultural catastrophe." In fact, the whole field of genetically engineered organisms—now including many essential crops that have been tinkered with to be herbicide-tolerant—could repeat the corn blight fiasco on a larger scale.

The potato makes another great study in how modern agriculture cultivates catastrophic vulnerability. The Andean tuber, now one of the world's most important sources of carbohydrates, is currently being battered by a remarkable oomycete (a funguslike plant related to brown algae) that changed the history of Ireland and is now once again melting spuds and the lives of the poor all around the world. Pat Mooney, a world-famous researcher on agricultural diversity, coolly describes *Phytophthora infestans* (which means "plant destroyer" in Greek) as a "hot zone" pest more deadly than Ebola: "It attacks the species *Homo sapiens* by cutting off its food supply." It now threatens to do to the globe's $40-billion potato industry what it did to the Irish in the 1840s.

Late potato blight originated under the Tolua volcano in Mexico thousands of years ago. There the oomycete, or "water mold," used to live an undistinguished life among its fungal cousins in the roots of cool humid forests. However, when the Spanish arrived, they introduced single-crop planting, which gave the blight a chance to sow some wild oats. It began its first global odyssey in 1843 when a curious botanist accidentally picked up a sample contaminated by blight and took it home to Philadelphia. Soon a mysterious disease started to eat spuds there. Traveling on seed potatoes, it later popped up in Belgium, where it found a parasite's heaven: small fields of genetically uniform potatoes everywhere.

By the mid-19th century, the potato had changed the face of Europe. Plucked from the Andes in the 17th century, the hardy spud, a cool-weather plant, had slowly colonized northern Europe in the 18th century. Peasants knew a good thing when they planted one: Potatoes could feed a family of five on just an acre and a half of land, a feat no other plant could match. So they adopted the Peruvian foreigner as insurance against blighted and rusty grain harvests, greedy landlords, and economic depressions. By the time blight arrived, the tuber had filled the cooking pots of peasants in Ireland, Scotland, Norway, France, Prussia, and Belgium and fueled an unprecedented population explosion.

The potato revolution (combined with increased acreage of another plant alien, corn) launched Europe's global expansion by staving off famine. Thanks to boiled potatoes, Europeans were able to multiply, industrialize, and generally make a nuisance of themselves around the world. In Ireland alone, the population grew from 2 million to nearly 8 million souls on the strength of the tuber in just 60 years. With the assistance of corn, the Andean wonder freed up land for small grains and vegetables. It eventually took Europe's population from 105 million in 1650 to 390 million in 1900. Europeans with bellies full of New World tubers or corn soon started to colonize Argentina, Canada, Australia, and New Zealand—setting off, of course, a string of other biological invasions. The historian Alfred Crosby, Jr., believes that the introduction of corn and potatoes to Europe kick-started one of the world's most dramatic demographic revolutions. "For the first time in human history a people were able to break out of the cycle of advance and retreat, of ascent and crash, a cycle that matched every era of success with an era of dismay."

The potato blight, however, began a counterrevolution. In Ireland it quickly erased all the exuberant human growth fed by spuds. Blight immediately robbed laborers and peasants of their sustenance: an average of 5.5 kilograms (12 pounds) of praties a day per person. Aided by wet weather, the blight decomposed entire fields in 24 hours and kept on doing so for four years in a row. The Great Potato Famine, assisted by typhus and dysentery, killed nearly a million people. With no praties in the cupboard, families dined on cats and dogs, brewed nettle soup, dug up the corpses of diseased livestock, and stripped beaches bare of dulse and carrageen. Not surprisingly the blight, combined with England's appallingly incompetent famine relief, prompted a remarkable exodus. Thanks to the plant destroyer, the United States acquired much cheap labor, a wallop of Catholicism, and eventually a basketball team called the Celtics.

After remolding history in Britain, North America, and Australia, the blight moved on and buried more peasants in the Balkans, China, and parts of Africa. After its dramatic predations firmly established the study of plant diseases (there were huge debates about the cause), the potato killer quieted down. But ancient organisms like *Phytophthora* don't retire; they merely evolve into smarter forms. Global trade revived the pathogen in the 1970s and gave it another global tour when Europe, struggling with drought, imported 25,000 tons of Mexican potatoes. A new, more aggressive strain of blight tagged along. After melting crops in the Netherlands and Germany, the pathogen hitched a ride on the seed potato trade to Japan, Rwanda, and Brazil.

In the 1990s, *Phytophthora infestans* even got a sexual facelift and acquired enhanced vigor. Two different mutating asexual strains met, united, and created a sexual being capable of swapping chemical-resistant genes. The first casualty was metalaxyl, the most widely used blight fungicide in the world. The new sexual strain of blight also developed another trick: Its offspring could now infect the soil for years and wait until wet weather before emerging. So what used to be a spectacularly erratic and sporadic problem dependent on live potato plants, says the Cornell University professor William Fry, became "more persistent and troublesome."

Since then the new and improved late blight has unsettled potato production around the world and forced emergency meetings among plant pathologists. Like the Irish episode, this outbreak has largely beset the poor, this time in Asia and Africa, where the potato, the world's fourth most important crop, has become a mainstay of food production. Many countries, such as Poland, Egypt, China, and Russia, are becoming more dependent on the potato than the Irish were. Late blight has reduced the vital potato harvest by 30 percent in the developing world while wealthy farmers in Europe and the United States have been forced to increase the use of toxic fungicides by 25 percent. In some countries, farmers can't grow potatoes without spraying every three days.

Meanwhile the plant destroyer has expanded its appetite to tomatoes, pears, and melons. Although new disease-resistant cultivars are being used in some countries, the blight continues to terrorize farmers who are using older varieties in Africa and Asia. The pathogen now has two entire Web sites devoted to its destructive migrations. At an Asian workshop on the blight in 2004, Chinese researchers dubbed *Phytophthora* "the most destructive disease of food crops in the world."

Pat Mooney, a Canadian maverick among plant researchers, has summed up the fiasco as another example of man-made agricultural stupidity: "The present threat is the product of overenthusiastic potato evangelism—backstopped by aggressive company marketing (of seed potatoes); foolish breeding strategies; and the reliance on a single commercial fungicide." Traditional plant scientists have simply called recent developments "alarming."

Cacao, the bean that made chocolate a global addiction, is ringing more alarm bells. Five hundred years ago the tree produced a vital currency and beverage for the Aztecs and the Maya. At one time a man could even buy a prostitute for 10 cacao beans. But with the ambitious help of 19th-century European empire builders, the plant left its Central American home, colonized the human diet, created a candy craze, and became a vagabond crop in tropical paradises around the globe.

The intrusion of chocolate monocultures in landscapes accustomed to rain-forest diversity predictably invited unparalleled disease invasions. The opportunists included witches'-broom (a fungal infection that shoots up a surreal spray of hair from branch tips and flowers) and other fungal troublemakers, such as black pod rot. Once undermined by fungi, cacao magnates abandoned their blighted investments only to set up shop in other jungles. Fungal invaders combined with rains, drought, and peasant insurrections to turn cacao into "the bean of contention." In 1975 an exasperated plant pathologist and cocoa disease expert confessed the obvious: "It is important to forestall possible pressure from pest and disease; there is no longer a new continent to which cocoa can be taken to escape from them."

Brazil, for example, used to be the world's number two cacao producer, harvesting 440,000 tons (400,000 tonnes) of beans a year. But witches'-broom swept that bounty away. It first ate up plantations in the Amazon, then traveled clear across the vast country. Broom, whose spores don't live long, shouldn't have moved that fast. Agricultural experts soon discovered the troubling explanation when they found a branch wired to a cacao tree. "Somebody from the Amazon region had deliberately introduced the pathogen into Bahia and essentially destroyed their crop," says the Australian plant pathologist David Guest. How long West Africa will remain the bean's favored ground is any fungus's guess. But chocolate, like most foodstuffs, is now just one bite away from a disease disaster.

So too is the banana, one of the world's most nutritious plants. It now feeds 400 million people a year in Africa and Asia and fuels a $12-billion export trade. But the ancient plant that has inspired songs, elected tyrants, and created its own bloody republics may be a dying cultivar altogether. Thanks to successive tinkering by farmers over thousands of years, the banana is also one of the world's dumbest and most defenseless crops. As largely seedless and sterile mutants, edible varieties can be reproduced only from cuttings. Almost every banana is a clone that can trace its lineage back to just a few bananas 10,000 years ago. The banana, in short, has not had any good sex for a long time. Without a genetic fix or some good old-fashioned vegetative friction, warns Emile Frison, one of the world's top banana experts, "we can be fairly confident of a drastic decline in banana production worldwide, and possibly the complete collapse of the banana as a major staple as well as export crop."

Most of the world's banana plants and plantains (there are a thousand varieties) are now more genetically uniform than U.S. corn crops in 1970. As a consequence, pests and fungal diseases eat about half of the world's crop. In the 1950s, fusarium wilt, or Panama disease, a fungus that attacks the banana from its roots, dispatched the world's most popular commercial variety: the Gros Michel. That banana produced big hardy bunches and tasted great but had one weakness: no defenses against the fungal opportunist. When fungicides failed to slay the new disease, banana companies imitated their cacao peers by bulldozing new rain-forest real estate until the wilt followed them. But before Harry Belafonte could sing "Six foot, seven foot, eight foot, bunch," Panama disease had wiped out the Gros Michel. Industry managed to avert disaster only by finding a variety in Vietnam that had been cultivated by the Duke of Devonshire. The Cavendish, a much more delicate creature than the Gros Michel, was nonetheless resistant to the Panama disease and soon restored the banana export trade.

But the Cavendish was no match for black sigatoka, another happy fungal rider of rain, wind, and irrigation water. First discovered in Fiji in 1963, black sigatoka has traveled around the world via its spores, which cling to infected leaves routinely used in packing crates. Compared with its less lethal cousin, yellow sigatoka, the black variety produces more spores and kills more leaf tissue by turning banana leaves yellow, brown, and black. A good dose of sigatoka can reduce yields by 50 percent and prematurely ripen the fruit.

To slow down sigatoka's advance, the industry began drenching banana plantations and banana workers with toxic fungicides up to 50 times a year. No other crop except the potato requires that kind of chemical onslaught. (One-third of the price of every commercial banana goes to fungicide applications.) This chemical warfare, in turn, has created an epidemic of leukemia and sterility among plantation workers in Latin America. It has also selected varieties of sigatoka that are more resistant and nasty.

By the 1990s, the banana was getting limper and limper. Sigatoka destroyed banana production in Costa Rica and the Amazon first. Then it aligned itself with a new Asian variant of Panama disease, known as

Race 4, and polished off 30 to 50 percent of plantain production in Nigeria. In Uganda—where the word for banana, *mtooke,* also means "food"—sigatoka ate 40 percent of the harvest. Race 4 hasn't reached Latin America yet, but when it does, the Cavendish will likely disappear from North American grocery shelves.

To keep the plant alive, breeders in Honduras have laboriously developed a few resistant hybrids. (It can take 18 months to cross-breed a banana, and there is no guarantee any seeds will be produced.) Small-scale farmers in Cuba, East Africa, and tropical America have readily adopted one new disease-resistant type even though it tastes like an apple. Meanwhile big business is counting on better fungicides or biotechnology to save the Cavendish from extinction. But the banana's travails once again underscore the monotonous dangers of monoculture. "If this can happen to the world's most popular fruit with all humanity as its witness," recently remarked the Canadian biologist Robert Alison, "imagine what could happen to more obscure, but no less useful plants whose fates are less publicized and open to public scrutiny."

The very staff of life, wheat, remains another shaky monoculture. Socrates, who understood the nature of famine, once noted that "no one can be a statesman who is entirely ignorant of the problems of wheat." But modern agriculture has made us all ignoramuses. Although we foolishly regard a thousand different kinds of bread as some sort of birthright, biological invaders have other designs.

Long before wheat overtook the world's grasslands, domesticated Europeans, and secured its status as the world's most import cereal crop, members of the fungus clan persistently blasted it again and again. Stem rust (just one of 300 species of *Puccinia graminis*) has been a particularly successful invader and the author of hundreds of famines. Rust takes its name from its appearance: It can decorate a stalk of wheat the same way rust discolors metal. Although native to the Near East, stem rust rarely bothered wild cereal populations there. Wheat's wild cousins simply developed a complex system of defenses that delayed the parasite's growth. Domestic wheat, however, lost these antifungal tricks. Every year as many as 6,000 varieties of rust, including *Puccinia,* gobble up $500 billion worth.

Puccinia graminis has always been a most formidable adversary. Blessed with an ability to shift its sporelike shape as many as five times during its life cycle, stem rust can produce as many as 2,000 spores per day for two to three weeks during the growing season. It can also travel as high as 10,000 metres (33,000 feet) as well as switch hosts. When not dining on wheat, the parasite can munch comfortably on the barberry bush, where it can overwinter and make new races of rust sexually. French farmers observed this deadly partnership years ago when they found that rust infestations on wheat fields were worse near barberries.

Rust has blasted every wheat-dependent culture with high-yielding history, from the ancient Egyptians to the Chinese. A scourge of rust may have forced Moses and his indomitable people to cross the Red Sea. Abetted by climate change, rust probably undid the Egyptians, who paid the pyramid builders with leavened bread. The ancient Greeks noticed that heavy dews resulted in rusty fields and feared them accordingly. Rust decayed Roman wheat crops so often that in 700 BC, bread eaters started to appeal to a newly minted rust god, Robigus, or his female form, Robigo, to avert crop losses. In the third week of April, farmers routinely sacrificed a red dog or red sheep to shield their wheat fields from the winds blasting from the east: "Spare our crops, O Robigus. Is it not enough that thou couldst harm them?" The poet Ovid, a bread lover, wrote odes to rust.

But wherever man took wheat, stem rust and the barberry bush (the wood made good tool handles and the berries, good jam) soon followed. When European immigrants proudly plowed up North America's great grasslands from Canada to Mexico for their favorite cereal, wheat, they had no idea they were also creating a *Puccinia* pathway complete with barberry bushes. The famous plant breeder and historian J.R. Harlan says that rust adored this monoculture the way bacteria take to agar in a petri dish: "We have created a gigantic epidemiological system, in which several species of *Puccinia* migrate yearly over 4,000 kilometers [2,500 miles] from a land where they cannot oversummer to a land where they cannot overwinter. The system must be included among the marvels of the biological world."

The marvel begins in northern Mexico, where spores can survive the mild winter to begin their journey northward against waving fields of amber. Carried by wind (stem rust spores originating in Africa have climbed up 10,000 metres (33,000 feet) to catch currents across the Indian Ocean and infect plants in Western Australia), they land on barberry and wheat alike, germinating, infecting, and creating more spores and all the while stealing the life from leaves and stems as they reproduce. In the early 19th century, some infestations wiped out 50 percent of the crop. Every time farmers tried a different variety of wheat partially resistant to rust, the parasite found a virulent gene capable of neutralizing it. In 1917 rust ate up a billion cubic metres (300 million bushels) of North American wheat, forcing the United States to declare a voluntary program of "two wheatless days."

After that scare, the U.S. government began a campaign to pull up barberry bushes throughout the West. Without its sex hotel, rust could not make any new pathogenic offspring. Within 15 years, civil servants and farmers had dutifully pulled up more than 18 million bushes in 13 states. Although the frequency of epidemics lessened, *Puccinia* continued to mow down wheat fields. In the 1950s, rust devoured up to half the wheat crop in North Dakota and Minnesota. Although current wheat crops have two or more genes resistant to common types of rust, the damage remains. The barberry campaign ended in the 1980s because of lack of funds, and if the bush were to reinvade the West, even in modest numbers, rust would acquire once again a host that allows the fungi to swap genes and become more resistant. Epidemics costing hundreds of millions of dollars and yield losses of up to 350 million cubic metres (100 million bushels) are but a berry away.

Although rust remains a formidable scourge around the world, it has recently been eclipsed by fusarium head blight or scab. This venerable and multitalented fungus, which first appeared in England more than a hundred years ago, has in the last decade decimated wheat and barley crops in more than 23 states and several Canadian provinces. Frustrated scientists complain that this fungal invader "presents a range of exasperating complexities rarely seen in any single plant disease." Fusarium not only reduces crop yields by shriveling kernels, but it also produces a deadly mycotoxin that makes infected grain too poisonous to be used for flour, malt, or livestock feed. (Ergot, a fungal toxin that grows on rye, terrorized Europe for centuries. Bread made with the infected grain caused hallucination and death from gangrene.)

Climate change, monoculture, and farming practices all seem to have prompted fusarium's rampages. A series of wet springs combined with no-till agriculture (a practice that leaves stalks in the ground) have given the fungus liberal opportunities to overwinter and attack crops in North America. In recent years, fusarium scab has reduced wheat yields by 50 percent and clipped barley yields by 37 percent. Since fusarium's explosive appearance in the 1990s, the amount of land devoted to wheat production in North America's breadbasket has decreased by 31 percent. The scab has kept farm incomes below the U.S. federal poverty level and prompted many cash-strapped grain growers to drop their health insurance if not sell the barn.

Fusarium, in fact, has become another global farm-eater: "A friend of mine quit last spring and told me that all that survived was a 20-year-old pickup," one American barley grower recently reflected. "He managed to pay off all of his bank loans but lost all of his land and machinery. Auctioneers are saying they never had so many inquiries." Six years of successive scab epidemics have accelerated the destruction of rural communities in Minnesota, North Dakota, and Manitoba and concentrated more power in the hands of big business.

Although largely ignored by the media, these grim food fables are now becoming a fearsome global chorus. Monocultures invite disease, and global trade gives the invading pathogens unlimited mileage. The solutions are not terribly complicated. True food security for any nation lies in eating what it grows and limiting food imports. Abandoning monoculture for greater crop diversity has always been on every disaster prevention menu. Some Mexican communities have kept disease and famine at bay by planting a dozen crops in small plots, or milpas. The method remains as sustainable today as it was 4,000 years ago. In recent years, the Chinese have stemmed the advance of rice blast, another fungal celebrity, by eschewing monocultures and simply interplanting two varieties of rice in a field. Similar practices in Africa have reduced the amount of food harvested by pests and pathogens by 30 percent. But all of these solutions imply local as opposed to fashionable global thinking.

So the bad news is this: Our food supply is about as secure as a Florida trailer park in a hurricane. According to the Cornell University ecologist David Pimentel, approximately 70,000 pests and plant diseases now consume half of the world's crop yield every year, despite the use of $35 billion worth of pesticides. But that's just one broken plate at the table. We are also completely dependent on a few plant species of dubious genetic integrity that have limited natural defenses against a growing horde of invaders. Without cheap fossil fuels and mountains of pesticides, we can't protect our oasis of plenty for long. And each day the vagaries of climate change make our agricultural systems more unstable. What we eat is increasingly being eaten by a crowd of foreign guests and gate-crashers. But we are setting the table, sending out the invitations, and even paying for the airfare. We have even added the possibility of engineered pests and plants running amok.

To many farmers, the pandemonium in their fields simply explains what happens when a society removes all culture from agriculture.

Wendell Berry, an American farmer and essayist, now charges that the globalization of food production might well become the 20th century's most ruinous legacy. A uniform Soviet-style model has nearly wiped out local traditions, local farmers, and local crop diversity with the bold lie that technique can conquer all the quirks of Mother Nature: "As soon as pests, parasites, diseases, climate fluctuations and extremes are left out, resistance to these things is also left out; and this resistance, in the soil and the lives that come from the soil, is what we call health. And so for total control we have given up health."

J.R. Harlan, who traveled to more than 80 countries to document the vanishing diversity of agricultural crops, noted a decade ago that more than three-quarters of our food now comes from cereals. We are eaters of tamed grasses, pure and simple. This explains why just five cereal crops—wheat, corn, rice, barley, and soybeans—now provide the world with most of its nutrition. "What would happen if one of these should fail?" warned Harlan, a generally pleasant optimist, in *The Living Fields*. His rude answer reflected a good appreciation of history: "It has happened in the past; it probably will happen in the future."

When bad weather combined with rust, rot, canker, blight, or other pests undoes dinner, we will probably do what our ancestors have always done: Eat the future. In the Old Kingdom of Egypt, when drought and pestilence blasted crops and reduced life along the Nile to a "sandbank of Hell," famished parents ate Hefat and Hormer. In the 14th century, when the rains would not stop and each seed of grain produced only two more in northern Europe, hungry parents ate Hansel and Gretel. And during the great famine of 1878, when human flesh hung in Chinese market stalls, fathers ate Ming and Wen, husbands ate wives, and children ate their parents.

What kind of civilization would knowingly sow such a crop?

SIX

Resurrecting Anthrax

"The Orwell law of the future:
any new technology that can *be tried* will *be."*
—David Gelernter

B ob Stevens, an affable photo retouching artist for *The National Enquirer* and other tabloids, didn't know much about anthrax, or woolsorter's disease, in the fall of 2001. Like most ordinary citizens, the Florida resident and Guinness drinker had no idea that U.S. and Russian scientists had "weaponized" the organism or that it had become the Lady Macbeth of biological warfare. He hadn't heard about the diabolical work of General Ishii Shiro and didn't know a damn thing about the world's largest anthrax outbreak, in Rhodesia in the late 1970s. Anthrax just wasn't a household name then. The victims of germ warfare rarely know their killers, let alone their secret origins or makers. Stevens was no exception.

But like a 19th-century woolsorter, the 63-year-old unexpectedly met the invaders in his workplace. A co-worker, Ernesto Blanco, the mailroom supervisor, innocently brought the agent right to his door. When Blanco passed Stevens's desk on his normal rounds sometime in the third week of September, a cryptic letter saying, "This Is Next. Take

Penicillin Now," leaked out its deadly contents throughout the American Media Incorporated building in Boca Raton, Florida. The poisonous letter, one of five sent to various media outlets on September 18, 2001, may have contaminated another package that came to the tabloid addressed to the sex goddess and singer Jennifer Lopez. Stevens, a JLo fan, had even retrieved that letter from the wastebasket to study its wacky contents, including some soap powder and a Star of David.

Anthrax is a shape-shifter, and the guise Stevens unhappily met was a spore. When in contact with the moist lymph glands of a sheep or human, anthrax assumes the form of rapidly reproducing bacterial cells. But once the bacterium has sucked all the energy from its host and killed it with a variety of toxins, the cells spill out of the remains. As soon as the germ encounters oxygen, it rapidly curls into a compact ball coated with tough protein. Inside this armor, spores can withstand fire, ice, poison, and time. They can go just about anywhere and endure almost any extreme event. For these reasons scientists have called anthrax "the perfect military pathogen" or "God's gift to warfare." About 200 anthrax spores can stretch across the width of a human hair, and a pinch of anthrax can contain a million spores. A killing dose can contain as few as 500 spores.

Nobody knows how many spores entered Stevens's lungs, but once there they stuck to his chest tissue like flies on sticky paper. The spores transmuted into rod-shaped cells that started making more *Bacillus anthracis*. Soon these bacterial colonizers started pumping toxins throughout Stevens's body. Within several days Stevens felt muscle pains and other flulike symptoms. In its early stages, anthrax mimics hundreds of other diseases.

On the fourth or fifth day, Stevens's breathing became labored. A throbbing headache and a nasty cough appeared. As the anthrax expropriated more and more of his body, Stevens felt worse. Respiratory distress, chest pains, and massive sweating followed. His last coherent words to his wife were "I love you." A couple of days later, he passed into a coma and became the first-known case of inhalational anthrax in the United States since 1976. At his autopsy the pathologists did rough calculations on how many spores were forming in Stevens's body and then discarded their contaminated tools in the man's corpse.

Much ado was made of the anthrax attack in 2001. It killed 5 Americans, injured 24 citizens (mostly postal workers), and frightened millions. While consumers temporarily elected Cipro the continent's most popular antibiotic, the FBI launched one of its biggest investigations ever, dubbed Amerithrax. The attack stopped the mail, revealed Grand Canyon–scale gaps in biological security, and ultimately cost the U.S. government tens of millions of dollars to decontaminate buildings hot with anthrax.

Nobody knows who sent the letters, but certain facts suggest a disturbing scenario: A government-trained scientist using a secret government recipe probably made the invader that killed Stevens. Everyone agrees that the "weaponized" anthrax that nearly paralyzed a nation for several weeks was the work of a professional. It contained finely powdered glass that gave the spores the "energy" to dance in the air in unseen clouds. As any bioweaponeer will tell you, an effective agent has to float like a butterfly and sting like a bee.

The strain employed, called Ames, had been cultured from a dead Texas cow in 1981 or earlier and had been a favorite of the U.S. military for decades. Only U.S. military contractors, scientists, and one vaccine manufacturer had access to the strain. "Thus the source of the attacks is thought to be domestic, someone with access to the military preparation or its secret manufacturing process," wrote the anthrax expert Meryl Nass in a damning paper in the *American Journal of Public Health*. Barbara Hatch Rosenberg, an equally combative molecular biologist at the State University of New York, also concluded that the very existence of a U.S. biodefense program made the anthrax attack possible.

America's bioweapons establishment certainly was the chief benefactor. After the attack the U.S. government allocated nearly $6 billion to bioweapons defense and research and breathed new life into the troubled anthrax vaccine industry. The number of laboratories and scientists handling dangerous germs has now mushroomed, "making oversight and adequate safety and security control much more difficult to impose," notes Rosenberg. Whatever claims scientists may make, biodefense has never been and will never be about protecting the public. Such a goal would be futile, given the variety and potency of engineered germs now available around the world. Biodefense is primarily about making germs in mass quantities to kill lots of civilians, explains Rosenberg. "Ultimately the very problem that made the anthrax attacks possible will be magnified." The attack, in other words, simply ensured that there would be more anthrax invasions and more anthrax makers in this world. The letter sender probably knew that.

The Amerithrax incident is just the latest stage in the remarkable evolution of anthrax, germ warfare, and man-made invaders. For nearly a hundred years, anthrax has been at the forefront of secret biological programs and science's quest to perfect a higher form of killing—the kind that leaves property intact and its owners dead. To this end, governments and their willing accomplices, scientists, have committed just about every known crime against the species and then some. The governments of Russia, the United States, the United Kingdom, Japan, and Canada have all deceived their citizens about the scale of their secret germ laboratories. In fact, no government that has ever cooked or weaponized anthrax has ever told the truth about its black work. Anthrax corrupts and weaponized anthrax corrupts absolutely.

Bacillus anthracis, the stuff of modern nightmares, is as ancient as a moonlit sky. It has probably occupied soils around the world for more than a billion years. Humans first encountered anthrax when they started hunting wild sheep more than 40,000 years ago. Killing and skinning sheep is a fairly mundane but messy task that can expose skin or lungs to anthrax spores. During the Bronze Age, anthrax probably became a constant companion of herdsmen, butchers, shearers, wool handlers, and tanners, and it has dogged these trades ever since.

The Greeks, who prized good animal husbandry, were among the first to record that the disease turned the skin of infected cattle, goats, and sheep black as coal dust. That's why anthrax takes its name from the Greek word for coal. They knew that livestock generally picked up the disease by eating grass or drinking contaminated water. (An anthrax spore can survive in water for up to two years.) And they observed that animals bled out at the nose and mouth after a short fever. They knew that an animal with anthrax could be standing one moment and drop dead the next. The ancients also knew that floods and droughts, both great spore-unsettling events, generally preceded outbreaks in sheep and shepherds alike.

Anthrax can cook up three distinct diseases in man or animals, but cutaneous anthrax remains the most common. It occurs when an anthrax spore invades a break in the skin, forms a nasty blister, and ulcerates the surrounding area. It is rarely a killer unless the infection invades the bloodstream. Gastrointestinal anthrax is a great animal dispatcher that starts when hungry herbivores gulp down spores. But it too can afflict humans. Many a wedding banquet in the Middle East or Russia has ended with grief when the party dined on meat infected by anthrax spores. The inhalation form of anthrax may be the rarest yet the most deadly. It has a mortality rate of more than 80 percent.

Cutaneous and gastrointestinal anthrax appear throughout history. The Old Testament of the Bible famously records one of the world's first cutaneous outbreaks. When Ramses II of Egypt ignored Moses' pleas to release the Israelites, God visited the pharoah's people with 10 plagues including fleas, boils, and frogs. The fifth plague specifically killed cattle, sheep, and goats and was probably anthrax. It was followed by a sixth that sprouted "festering boils" on man and beast, another anthrax trade-mark. God's instructions to Moses read like a modern bioterror manual.

God asked Moses to take handfuls of soot from a furnace and toss the stuff into the air so it would make a fine dust over Egypt. Moses did just that and cutaneous anthrax quickly erupted. Moses may have been the world's first bioweaponeer.

Nearly 2,000 years ago, the Roman poet Virgil described an anthrax epidemic among sheep with even greater accuracy: "The pelts of the diseased animals were useless and neither water nor fire could cleanse the taint from their flesh…. If anyone wore a garment made from tainted wool, his limbs were soon attacked by inflamed blisters and a foul exudate, and if he delayed too long to remove the material, a violent inflammation consumed the parts it had touched." This was classic anthrax.

As cattle and sheep herds grew in size during the Middle Ages, anthrax earned a fearful reputation throughout Europe. Peasants called it the "black bane." It repeatedly killed entire herds and precipitated famines. The French called it *charbon,* which means "coal"; villagers would drag off individuals afflicted with skin sores and isolate them in far-off places. During the Black Death in the 13th century, a period of successive waves of mass dying because of plague and other pestilence, anthrax most assuredly carried one of the scythes.

By the 18th century, anthrax had become such a serious drain on economies that several prominent aristocrats started to study it. Their detailed investigations gave birth to veterinary medicine and took on new intensity after 1769, when an outbreak of anthrax on Saint-Domingue (now Haiti) killed thousands of cattle and caused 15,000 human deaths from gastrointestinal anthrax. In successive epidemics, slaves who butchered meat, tanned skins, or performed animal enemas all contracted black lesions on their skin. Long before the German scientist Robert Koch tied a specific germ to anthrax, the French had largely sketched out the bacillus's natural history and modes of transmission.

Anthrax became such a prominent subject in Europe that the famous 18th-century physician Jean Astruc even included the disease in his classic treatise on syphilis. Astruc believed that syphilis was a modern affair and not a disappointment shared by his European ancestors. Old tales about young drunkards who grew carbuncles on their penises after bedding the fair sex ("in a Fortnight's Time his private Parts mortified and dropped off of their own accord") weren't about syphilis at all, argued Astruc. He suspected the carbuncle was anthrax, and he might have been right. Probably more than one rural couple using a contaminated sheep's wool or leather hide for their trysts could have wound up with a bad case of genital ulceration caused by anthrax.

In the 19th century, Louis Pasteur, a man dismissed as a "charlatan chemist" by many of his peers, solved one of anthrax's great riddles. For centuries peasants had spoken of fields "cursed with anthrax." Any animal that grazed in such a place came down with the bane. Pasteur found the answer while visiting a sheep farmer who had lost several sheep to *charbon*. On a tour of his farm the scientist noted that the color of the earth looked different in one spot. The farmer explained that was where he had buried his sheep felled by anthrax. On closer inspection Pasteur found numerous worm castings on top of the soil. He guessed that the worms were bringing anthrax spores to the surface, where the curse was then ingested by the sheep. He then proved his point by inoculating guinea pigs with worm dung. They all came down with anthrax. One of Pasteur's students later proved that fertilizer made from dead animals moved the curse around Europe more efficiently than earthworms did.

Anthrax next made headlines when imported alpaca and mohair from Peru and Asia Minor sickened scores of British woolsorters in the late 19th century. After inhaling dry and dusty imported wool laced with bacilli, many workers died 12 to 17 hours later, with nary a cough nor a purging. But John Bell, a crusading London doctor who eventually improved working conditions for sorters, found the blood of the dead completely invaded; it abounded "with bacteria-like caterpillars with legs—millions in a drop."

By the end of the 19th century, anthrax had become a singular scientific wonder. Nature's hardiest and most durable bacterium served as biology's first window on the secrets of the microbial world. Thanks to the pioneering labors of Pasteur and Koch, anthrax became the first germ to be conclusively linked to human disease, the first organism to prove the "germ theory," the first to legitimize the new field of microbiology, and the first organism to be used to produce a live bacterial vaccine. With all these firsts, it's not surprising that scientists soon discovered that the same tools used to discover, describe, and treat an anthrax infection could also be turned to diabolical deeds. And so microbiology quickly gave birth to black biology, and anthrax became its dominant obsession.

The Germans, Europe's most technically advanced people at the turn of the 20th century, could claim to be the first to use anthrax as a weapon during World War I. Not satisfied with the killing power of mustard gas, the kaiser's ambitious disciples in the army's veterinary corps started playing with bacteria. Given the military's dependence on livestock for transport, the veterinarians knew that they could stop an army dead by sickening its horses and mules. Glanders, a devastating horse feller, and anthrax proved to be the best candidates and were easy to culture in a lab.

Compared with the Amerithrax attack, the Germans' first assaults with anthrax were simple. Across France and in Bucharest, German agents smeared vials of the glanders bacterium on the gums of cattle and horses. Spies armed with bottles of liquid anthrax and cork-tipped needles sought out military horses stabled in New York. Dr. Anton Dilger made a batch of anthrax spores in a secret laboratory in Maryland for use on draft animals in Baltimore. In Argentina more secret agents fed animals bound for the Western Front with sugar cubes laced with anthrax. British documents indicate that the Germans infected nearly 5,000 mules and horses in Mesopotamia alone and tried to infect sheep bound for Russia. Norwegians even arrested one German bioweaponeer, a baron, trying to infect draft reindeer with anthrax.

These crude trials, however, were soon outdone by General Ishii Shiro, a Kyoto immunologist with a keen interest in self-aggrandizement and medical immortality. The general occupies a dreadful chair as the father of biological warfare. After a two-year tour of the world's top bacteriological research facilities in the late 1920s, Ishii concluded that biological warfare was the way of the future. For starters, it not only was cheaper than conventional warfare but also possessed "distinct possibilities." Why else would the League of Nations have banned it in 1925, reasoned Ishii. He quickly deduced that there were two kinds of biological warfare research: Type B, or defensive, which involved vaccines; and Type A, or offensive, which involved field trials with human guinea pigs. Type A, of course, could be done only overseas. Recently invaded Manchuria made a perfect laboratory.

With the blessing of the Japanese military, Ishii set up his first biological warfare program in Harbin, capital of Heilongjiang province, in 1932. At the time, Harbin was a global village that boasted a mix of Jews, White Russians, American adventurers, and European confidence men. His troops occupied two blocks in the industrial part of the town and put up watchtowers and barbed wire. In the nearby town of Beiyinhe, Ishii got to work using captured bandits, Communist rebels, and guerrillas caught harassing Japanese troops. The general disguised his secret experimentation site as a routine military prison: the Zhong Ma Camp. He fed his victims well and took blood samples regularly. Then he exposed them to plague, anthrax, potassium cyanide, and phosgene gas. After often live dissections without anesthetics (the doctors didn't want to spoil any tissue), Ishii burned what he called the "sacrifices" in a crematorium. It didn't take long to prove to his superiors that germs could kill lots of people.

But the good doctor had grander plans for the emperor. In 1939 he constructed the world's largest biological warfare factory with the help of 15,000 Chinese laborers in Ping Fan. The facility, blessed with the Orwellian title of Water Purification Unit 731, occupied an area larger than Auschwitz-Birkenau and consisted of 76 buildings, including three furnaces for the dead and a brothel full of "comfort women" for weary microbiologists. At its peak capacity, the army of germ cultivators at Ping Fan turned out 33 kilograms (72 pounds) of plague monthly and nearly 600 kilograms (1,300 pounds) of anthrax.

Ishii typically prepared squeamish medical recruits for the tough work with a jaunty pep talk: "I beseech you to pursue this research based on the dual thrill of a scientist to exert efforts to probing for the truth in natural science and research into, and discovery of, the unknown world, and, as a military person, to successfully build a powerful military weapon against the enemy." After building a lumber mill on the site, the Japanese referred to their prisoners as *maruta,* or logs.

To discover the unknown world of anthrax, Ishii's unit routinely tied logs or "experimental materials" to the ground and then exploded anthrax bombs upwind. Some of the logs wore helmets and body armor so they wouldn't die from shrapnel wounds first. Doctors then watched for signs of disease by the hour. One hardworking anthrax innovator noted in a 406-page dissection report on 43 logs that "nine cases were infected perorally with some foodstuffs which contain some quantity of anthrax bacillus and all patients died definitely after several days by acute abdominal symptoms and severe hemorrhagic ascites." The logs didn't always sit still for their germ attacks. When 40 prisoners broke loose in 1944 at the Anda airfield, where Ping Fan's anthrax entrepreneurs usually did their field trials, exasperated doctors simply got behind the wheels of trucks and ran them down, one by one.

When not experimenting with man-made anthrax, Ishii harnessed the possibilities of other biological invaders. His unit released plague-infested fleas in several Chinese communities and then carefully counted the dead. His men dumped typhoid bacilli into ponds, wells, and reservoirs and then took people's temperatures. Unit 731 dropped porcelain bombs containing infected fleas on bucolic villages and then checked for the inevitable fallout. Ishii's curiosity knew no bounds. He ordered his troops to inoculate peasants with cholera and even distributed chocolates laced with anthrax to Chinese children.

The American historian Sheldon Harris calculates that Ishii constructed as many as 26 research facilities in China, exploring every possible variety of death. All of the factories used humans for field trials, and each one exposed human prisoners and animals to every bacteria, virus, and poison then known to science. At a factory called the Anti-epizootic Protection of Horses Unit, in Changchun, some of Japan's foremost vets, botanists, and pathologists studied killing properties of glanders, plague, and anthrax. After exposing "experimental materials" to a particular pathogen, the professionals then executed them because "the materials were no longer worth keeping for further experiments." The scientists did such a thorough job of contaminating the neighborhood that local Chinese were contracting plague and anthrax long after the war.

While Ishii made a mockery of the Hippocratic oath, anthrax caught the interest of a gregarious Canadian, Sir Frederick Banting, one of the discovers of insulin. In 1937 the small-town physician started to worry about bacteriological warfare. While many of his peers moped about the hazards of chemical warfare (which Banting had seen at first hand in the trenches), the doctor suspected that bacteria could be an even greater threat in "a war of scientist against scientist." Unlike many of his stodgy British colleagues who didn't take biological warfare seriously, Banting could see the future. He was convinced that airplanes could be used to spread deadly diseases, such as typhoid and cholera. He also had no doubt that anthrax would be used to wipe out herds of domestic animals. In the event of a new war, Germany, Italy, and Japan, he reasoned, would use bacterial agents in the blink of a eye.

In a prescient 19-page memo to Canada's National Research Council, Banting imagined all the horrors that Ishii had already realized: infected bombs, aerial spraying of civilian populations, and the rearing of disease-carrying insects. War used to require 10 people at home working to support each soldier abroad, he wrote. But bacteriological warfare changed all the math: "It is just as effective to kill or disable ten unarmed workers at home as to put a soldier out of action and if this can be done with less risk, then it would be advantageous to employ any mode of warfare to accomplish this."

Three years later Banting found himself head of the Western world's first bacteriological warfare unit, funded almost entirely by prominent Canadian capitalists. He hired a number of bacteriologists and veterinarians who toyed with military applications of parrot fever and rinderpest. Their first aerial trial involved flying a plane over the cottage country north of Toronto, where they dropped uninfected sawdust particles out the door. Some grades of sawdust traveled farther than others, they observed. Meanwhile in China, Ishii was dropping plague-ridden fleas out of planes in porcelain bombs, an innovation that set off several plague epidemics and killed thousands of Chinese citizens.

By now Banting had been infected by the cruel and awful nature of his work. Black biology, it seems, invites blacker thoughts. At night he dreamed of killing "3 or 4-million huns" and watching them "wriggle and stew in their own juice." He requested a factory to make bacteria in mass quantities so that Canada could retaliate if the Germans ever used germ warfare. (Hitler, a victim of gassing during World War I, didn't support biological warfare, but many Nazi scientists ignored his prohibitions and conducted Ishii-like experiments in the concentration camps.)

Banting died in a plane crash in 1940. The government eventually found a fitting locale for a factory on an island downstream from Quebec City. In the 19th century Grosse-Île had served as a quarantine station; more than 12,000 Irish immigrants died there of cholera. But before the Canadians started to produce batches of anthrax on the graves of cholera victims, the British, latecomers to germ warfare, eagerly contaminated a 210-hectare (520-acre) island in northwest Scotland.

Gruinard Island, a good home for sheep, lies off the west coast, halfway between the pretty town of Ullapool and Gairloch in the

Highlands. In 1941 the British government abruptly expropriated the place, expelled the lone shepherd, and renamed it "X Base." Fifty scientists and their hapless helpers then invaded Gruinard to test the feasibility of using anthrax (code-named N) in bombs and bullets. The British, like the Japanese, appreciated the finer killing properties of anthrax and even mass-produced a batch of anthrax-laced cattle biscuits in a project called Operation Vegetarian.

On Gruinard the scientists employed straightforward methods. The N pioneers erected a wooden gantry from which they suspended bombs hand-filled with anthrax spores. From a distance, hooded scientists detonated the bombs 1.2 meters (4 feet) above the ground. Unsuspecting sheep downwind then calmly breathed clouds of anthrax spores the same way Ishii's unsuspecting *maruta* did at Ping Fan.

Although the scientists calculated that the explosion might actually kill most of the spores, they discovered that a spore survival rate of just 10 percent did the job. Three days after the detonation, the animals fell sick and started to bleed from the nose. In fact, the first test killed 90 percent of the exposed sheep. Their bodies were thrown over a cliff and buried under a dynamited rock slide. These open-air tests also spread spores over much of the island, as did subsequent trials.

When anthrax broke out among sheep on the mainland the next year, Britain's anthrax pioneers realized that bioterror had its shortcomings. To contain the spread of anthrax, they immediately set the island's heather on fire. The following year they tested the soil again only to find that anthrax had easily survived the firestorm. Next the government posted signs prohibiting boat landings, saying, "This island is government property under experiment." In a final report on the Gruinard experiments, scientists excitedly noted that anthrax could make cities "uninhabitable" for generations.

Indeed, 43 years later anthrax still ruled over Gruinard. In 1986 a team dressed in space suits returned to soak 4 hectares (10 acres) of ground with a mixture of 2,000 tonnes (2,200 tons) of seawater and

280 tonnes (310 tons) of formaldehyde, a true spore killer. But "Anthrax Island" didn't get a clean bill of health until 1990. Although sheep once again graze the island, archeologists, who have found 400-year-old anthrax spores in the graveyards of medieval hospitals, swear they wouldn't walk there.

After receiving worrisome reports about Japanese experiments in China from missionaries and intelligence officers, the Americans belatedly got into the anthrax business by setting up a biological warfare unit at Camp Detrick, Maryland, in 1943. The government called it "the Institute"; scientists dubbed their first lab "Black Maria." The technicians had a simple charge: "Get the germs to the enemy forces as intact and deadly as when they began." Although the scientists tinkered with dysentery, glanders, cholera, and brucellosis (another cattle killer), anthrax proved to be the most "professional pathogen."

To test their handiwork, the Americans established a field site near Granite Peak, Utah, known as the Dugway Proving Grounds. In Vigo, Indiana, they constructed facilities to manufacture 50,000 anthrax bombs a month. A pilot program eventually mass-produced anthrax from a 38,000-liter (10,000-gallon) tank. The British ordered 500,000 bomblets, each capable of releasing a handful of anthrax. At its height the Camp Detrick program employed more scientists than Ishii's factory of death at Ping Fan. "We were fighting a fire and it seemed necessary to risk getting dirty as well as burnt," the well-known science writer Theodor Rosebury reflected in 1963.

Anthrax got another boost in 1945 when Russian troops invaded Manchuria and discovered the remains of one of Ishii's monstrous death factories. Captured documents and Japanese prisoners detailed Ishii's experiments with anthrax, dysentery, and cholera on "experimental materials." The news didn't surprise the Soviets any more than it later did the Americans. Ishii's boldness, however, did inspire Soviet technicians.

Ever since World War I, the Bolsheviks had taken a keen interest in biological warfare thanks to a typhus epidemic that had dispatched nearly 10 million citizens between 1918 and 1921. Long an army follower and an opportunist of economic mayhem, typhus travels in lice and kills 40 percent of the afflicted with high fever and gangrene. Impressed by typhus's killing abilities, the Soviets decided to put it to work. By 1928 they had even set up a lab to grow the germ by using chicken embryos.

By the mid-1930s, the Soviet Union's biological warfare program had copied Ishii's projects. Scientists occupied Solovetsky Island in the White Sea, where they began working with Q fever, melioidosis (a debilitating tropical infection that later dogged U.S. troops in Vietnam), and glanders. Like the Japanese, the Soviets quietly experimented on prisoners and scientists alike. Tests often ended with short notes: "This experiment was not finished due to the death of the researcher."

The Soviets also weren't shy about deploying crude bacterial weapons in their life-and-death struggle with the Germans. Tularemia, or rabbit fever, seemed an ideal candidate: It didn't kill but it did immobilize the infected with pneumonia for a long time. To disperse this agent, the Soviets probably used crop sprayers and as a result encountered a bad case of what the profession calls "boomerang" or "blowback." The statistics on tularemia tell the story. In 1941 Soviets endured about 10,000 natural cases of rabbit fever. But in 1942 the caseload inexplicably jumped to 100,000. That same year a curious epidemic stopped an advance of German troops in the Volga region near Stalingrad. But thousands of local troops and civilians also got sick. When Ken Alibek (who later directed a Soviet bioweapons program) raised these facts with his military instructor in 1973, the answer was swift and dismissive: "I want you to do me a favor and forget you ever said what you just did. I will forget it too."

In 1945 the captured documents on Japan's secret program awoke many memories and gave the Soviets brighter and bigger ideas. For starters they illustrated the advantages of producing biological weapons the way an industry builds cars. The Soviets even copied Ishii's blueprints while building a germ factory in Sverdlovsk. Following the general's lead, they started to research crop killers and livestock cullers and sweetly called the whole program "Ecology." After experimenting with foot-and-mouth and psittacosis, the Soviets, like the Japanese, found anthrax.

Meanwhile the Americans had finally found Ishii. After wading through months of denials, evasions, and deceptions from various members of the Japanese military, naive U.S. investigators discovered the commander of Unit 731 happily ensconced in his home, feigning illness, in January 1946. Ishii dodged every moral question and played with the Americans for more than a year. He said the emperor didn't know a thing about biological warfare because "the emperor is a lover of humanity and never would have consented to such a thing." He described Japan's activities as small, local, and defensive.

The bioweaponeer assessed the mind-set of his captors with stunning accuracy and concluded that his knowledge of human experiments ("the Secret of Secrets") could be bartered for immunity from war crimes. "If you will give documentary immunity for myself, superiors, and subordinates, I can get all the information for you," he told one startled interrogator.

The Americans didn't ponder the possibilities for long before making a bargain with the devil. Camp Detrick's scientists hungered for the world's first data on biological field trials involving humans and Ishii had it, complete with slides and reports on live dissections. At the time one eager scientist gloated that the information "may prove invaluable." Another explained, "Such information could not be obtained in our own laboratories because of scruples attached to human experimentation." So the Tokyo war trials came and went without any mention of Ishii or his Secret of Secrets. The deaths of thousands of human guinea pigs conveniently disappeared from the public record in Japan and the

United States. The ever-practical U.S. military understood that it couldn't try Ishii for war crimes without spilling the beans on its own expertise as well.

To this day the U.S. government has not disclosed the full cover-up of Ishii's human experiments nor the scientific community's appetite for the gruesome revelations of his work. In 2001 the *Journal of the American Medical Association* even refused to publish a series of articles by Sheldon Harris on Japanese research and U.S. complicity after the war. Medical reviewers said the charges were just too "sweeping and severe." Anthrax won again.

After the war both the Americans and the Soviets ramped up their germ factories. The star product remained weaponized anthrax that floated more lightly than talcum powder. At Camp Detrick, engineers constructed a four-story chamber nicknamed the "eight ball" solely for anthrax tests. They placed donkeys and monkeys inside, exploded anthrax bombs, and then took notes. They constructed fermenters to cook up large batches of anthrax on a scale matching Ping Fan's. So much anthrax was being cooked and studied that accidents happened all the time. Whenever scientists died from infections, the brass at Camp Detrick buried the truth seekers in lead caskets and listed their deaths as "occupational death from respiratory disease." Sometimes they even harvested the anthrax from the dead, because anthrax bacilli get more virulent or useful to bioweaponeers after they pass through a host. After William Boyles died from anthrax exposure, the strain of the germ used at Detrick changed from Vollum to Vollum 1B. The B stood for Boyles.

U.S. scientists never got as callous as the Japanese. In fact, they solved some of the ethical problems associated with human testing by recruiting conscientious objectors from the Seventh-Day Adventists Church. For Operation Whitecoat more than 2,300 Adventists voluntarily breathed aerosol versions of Q fever, typhus, and tularemia. Treatment was often withheld for as long as possible to measure the full effect of the aerosols. The government told volunteers for the program between 1954 and 1973 that the studies were for "defensive" purposes only.

But having individuals inhale air samples wasn't sufficient. The U.S. military needed information on whole populations and secretly went to its citizens for large open-air vulnerability tests. But instead of scattering anthrax over American neighborhoods, the military prudently used a variety of "harmless" surrogate bacteria or simulants "under the safest and most controlled conditions possible." The surrogates included *Bacillus subtilis* (also known as *Bacillus globigii*), a close relative of anthrax that exists in spore form; *Serratia marcescens,* a bacterium that leaves a traceable red pigment in its wake; and *Aspergillus fumigatus,* an opportunistic fungus. None of these surrogates, however, is truly harmless. *Bacillus subtilis* has the ability to cause trouble among the immunocompromised and is genetically unstable. *Serratia marcescens* can trigger a bad case of meningitis and even arthritis. And *Aspergillus* had a nasty reputation even in 1949 as a "contaminant of lesions and ... agent of infection" among the sick and infirm.

But the siren call of scientific progress once again proved stronger than oaths of "First do no harm" or civil liberties such as informed consent. At a trial in a lawsuit about open-air testing on civilians in 1981, one general admitted that the government simply didn't have a choice. In order to gauge the full impact of biological weapons, "You have to test them in the kind of place where people live and work."

To their credit, the military scientists started at home: They secretly released *Serratia* in the Pentagon's air vents. The test illustrated that an anthrax attack would knock out most of the top brass. CIA agents then drove around New York in a car with a modified muffler spewing out surrogate germs. Off the coasts of Antigua, Nevis, and St. Kitts, the British also got into the act. The Royal Navy bombarded floating arks of monkeys and guinea pigs with hot agents, including anthrax, tularemia, and brucellosis.

The bioweaponeers also left their bacterial calling card in San Francisco. In 1950 the army conducted six biological attacks on the city by whipping up aerosols aboard ships off the coast. In a study of "the offensive possibilities of attacking a seaport city with a BW aerosol from offshore," the scientists released *Bacillus subtilis* and *Serratia marcescens* in such hearty clouds that nearly every citizen of San Francisco inhaled millions of bacteria. One cloud traveled as far inland as 42 kilometers (26 miles). "Had you been in that area, I assure you, you would have gotten sick had this *Bacillus globigii* been anthrax," boasted William Patrick 40 years later. Patrick, now a bio-threats consultant, then worked as the director of Camp Detrick's product development division.

The open-air test did make many people sick. A few days after the test, alarmed physicians described a baffling epidemic of *Serratia marcescens* infection that sickened 11 hospital patients and killed a retired pipe fitter, Edward Nevin. The public didn't learn about the tests or the infections until Nevin's family sued the military in 1980.

In 1952 the same experiment was repeated in Alabama, with similar appalling results. After the army sprayed Florida's Gulf Coast with *Hemophilus pertussis,* the incidence of whooping cough there rose from 339 cases and 1 death in 1954 to 1,080 infections and 12 deaths in 1955. To establish the "behavior of aerosol clouds within cities" the bioweaponeers bombarded the cities of St. Louis, Minneapolis, and Winnipeg with repeated aerial sprayings. In size and climate these unfortunate urban centers looked a lot like target Soviet cities.

Some of the experiments defied belief. In one case bioweaponeers brazenly exposed travelers at Washington's National Airport to "harmless" germs in 1956, using secret-agent-style suitcases equipped with aerosol generators. The testers concluded that for some citizens, "the calculated exposures would have been massive doses if that many pathogenic organisms had been inhaled."

In 1965 Fort Detrick's bioweaponeers got bolder and invaded New York. To establish the vulnerability of subway passengers to "covert attacks with biological agents," they exposed more than a million New Yorkers to *Bacillus subtilis*. (The Soviets later repeated the very same experiment on the Moscow subway system, proving that dark minds think alike.) Armed with fake ID, the researchers dropped lightbulbs, each containing 87 trillion bacilli of the anthrax substitute, onto ventilation grilles. Then they coolly monitored air currents and dissemination patterns.

The collected data showed that moving trains sucked the "agent aerosol" along the entire system. Not surprisingly, the army concluded that the lightbulb attack was a hell of a good way to infect the working population of downtown New York "at a period of peak traffic." William Patrick recalled, "We would have infected 500,000, and that's conservative thinking." No one tallied the number of illnesses and deaths caused by the tests because health monitoring wasn't part of the study. But as Patrick put it: "We clobbered the Lexington line."

To establish how large an area a biological agent could effectively contaminate, the army also employed a cancerous chemical called zinc cadmium sulfide, an easily traceable compound. Cadmium is a cancer maker and a longtime staple of chemical warfare. Areas in Ohio, Kentucky, and Texas all got sprayed, as did the entire city of Winnipeg.

All in all, between 1949 and 1969 the U.S. military secretly bombarded citizens with bacteria and bacterial surrogates in 239 separate open-air tests. In 1970 President Richard Nixon officially ended the offensive germs program in an attempt to save money and discourage any more madness. "Mankind already carries in its hands too many seeds of its own destruction," he said.

The Soviets continued to sow their own anthrax seeds. By 1972 they had built the world's largest bioweapons program, called Biopreparat. It employed nearly 60,000 scientists and technicians and consisted of scores of factories and laboratories as well as a testing ground on Rebirth Island in the Aral Sea. Anthrax was its central obsession until the 1990s, when the Soviets graduated to smallpox and weapons that combined more than one disease.

Anthrax made a natural choice for Soviet bioweaponeers. Known as the "Siberian ulcer," anthrax had slowed down the country's relentless colonization of the north since the 19th century. Every time the czar or a commissar pushed people and their grazing animals further north, anthrax just pushed back. In some years as many as 50,000 animals died. At any given time, as many as 12,000 peasants suffered from the Siberian ulcer.

Throughout the 1950s, the Soviets played with anthrax in an attempt to coax out a "battle strain," a variety that was deadly and mobile and could be produced in bulk. At one facility some liquid anthrax found its way into a city's sewer system, where it mixed with the rodent population. The army disinfected the sewers but anthrax survived among the rats, where it evolved into an even more deadly strain. When a bioweaponeer captured a local rodent and identified anthrax 836, the new strain became a candidate for mass production.

In the 1970s, Biopreparat's chief anthrax factory lay in the center of the country at Sverdlovsk, a gray industrial town that churned out tanks, bombs, and anthrax 836. At Compound 19, workers on three shifts manufactured a dry form of the germ for Soviet bombs and warheads. Sometime in early April 1979, a worker removed a clogged air filter on an exhaust pipe without telling the night shift. When the stressful business of drying cultures and grinding fine powders resumed, hours passed before anyone noticed the filter's absence. Meanwhile much of the working-class area downwind of Compound 19 had been exposed to weapons-grade anthrax. Within days the first cases were clogging the hospital. The flow of patients continued for more than a month as cleanup crews scrubbed trees and hosed down roofs, spreading the spores further. Emergency teams vaccinated thousands and handed out antibiotics to hundreds. In the end, as many as 105 people died of inhalational anthrax.

To avoid an international incident, Soviet spin doctors quickly blamed the outbreak on a truckload of contaminated meat from "unofficial" vendors. Rumors of the accident and the mystery epidemic didn't reach the West for another year. Although President Boris Yeltsin admitted in 1992 that "military developments" caused the outbreaks, Russian newspapers still blame the disaster on tainted food.

While the Soviets tried to cover their anthrax tracks, bioweaponeers used a messy colonial war in Rhodesia to conceal the largest outbreak of anthrax ever recorded among livestock and people. Between 1978 and 1980, an inexplicable explosion of anthrax killed thousands of cattle and infected more than 10,000 black farmers, killing 183. In an independent 1992 investigation of the suspicious outbreak, the American physician Meryl Nass concluded that the bacteria had been deliberately spread by Rhodesian troops in an attempt to poison the rural food supply of black guerrillas.

Her detective work produced a number of startling findings. At the time of the anthrax sortie, Rhodesia (now Zimbabwe) had one of the lowest incidences of anthrax in Africa. It had reported only 300 human cases between 1950 and 1978. The plague erupted in places where anthrax had never appeared before, moved into six of eight provinces (natural outbreaks always remain local), and was largely confined to tribal trust lands, where guerrillas depended on farmers for food.

Last but not least, the plague corresponded with the final months of a dirty guerrilla war in which white troops even poisoned guerrillas by abandoning denim clothing impregnated with rat killer and pesticides. Several years after Nass's investigation, former senior members of the Southern Rhodesian Security Forces admitted to using anthrax spores obtained from South African bioweaponeers "to kill off the cattle of tribesmen assisting guerrillas" and to demoralize rural blacks by sickening or killing their cattle. They also confessed to seeding rivers with cholera.

Back in the Soviet Union, Biopreparat moved its anthrax facilities to Stepnogorsk, in the northern deserts of Kazakhstan. The job of making a more dangerous strain of anthrax fell to an enterprising microbiologist, Ken Alibek. Using machines identical to those employed by soft-drink bottling plants, Alibek got to work. By 1987 Alibek and his crew had cooked up "the most effective anthrax weapon ever produced." In powder and liquid form, the new strain was three times stronger than the stuff that had escaped from Sverdlovsk.

In 1988 a group of colonels from the biological group of the Soviet army sat down with Alibek, now the first deputy chief of Biopreparat. The colonels proposed to cram giant SS-18 missiles with anthrax. Each missile carried ten 500-kilotonne warheads and could travel up to 9,600 kilometers (6,000 miles). Alibek knew that 100 kilograms (220 pounds) of anthrax could kill up to 3 million people and that just one SS-18 missile could finish off New York. With a few quick calculations, he figured that Biopreparat could deliver about 600 kilograms (1,300 pounds) for the colonels. In his fascinating exposé of the Soviet anthrax machine, *Biohazard,* Alibek recalls not "giving a moment's thought to the fact that we had just sketched out a plan to kill millions of people."

As the Soviets perfected weapons of mass destruction, the technology to make more anthrax weapons started going global. By the 1980s, India, Israel, Pakistan, China, North Korea, Taiwan, Vietnam, France, Libya, and South Africa had all entered the anthrax race. So too had Iraq. The Iraqis purchased germs from a biological supply company in Maryland and pharmaceutical equipment from Europe. They studied Ishii's experiments and probably added some grim improvements. By 1990 Saddam Hussein's eight biological warfare complexes had made or stored 8,350 liters (2,200 gallons) of anthrax as well as smaller quantities of botulinum, aflatoxin, and wheat smut. Analysts feared the Iraqi leader would use anthrax during the Gulf War, but he never did. Although United Nations weapons inspectors found and destroyed most of this equipment by 1996, its makers and administrators still walk the sands of the Middle East with anthrax on their minds.

In the 1990s, the Soviets tired of the Siberian ulcer and moved on to smallpox. Using cultures from India, scientists grew the virus on the kidney cells of monkeys in big reactors. One bioreactor alone produced enough of the virus to infect everyone on the planet more than 2,000 times, a remarkable example of overachievement. By the time the Soviets shut down the world's largest biological warfare program, Biopreparat had become a ripe pomegranate ready to spread its seeds. With the dissolution of the Soviet Union, its scientists dispersed to the far corners of the earth or to employers with deep pockets. They took with them their secrets and tricks for managing anthrax and smallpox. So anthrax and its makers did not disappear; they merely decentralized.

Anthrax continues to curse science the same way it cursed fields in medieval Europe. In Japan many members of Ishii's Unit 731 went on to long and prominent careers in that country's National Institute of Health (JNIH) and major pharmaceutical companies. Most, however, had trouble repressing their old habits. In the 1950s, Ishii's former colleagues injected inmates of a Tokyo prison with *Rickettsia typhi* and gave babies pathogenic *E. coli.* They also inoculated elementary students with untested vaccines and withheld polio vaccine during an epidemic. In the 1960s, Ishii's scientists refused to abandon a program of compulsory smallpox vaccination (long after the elimination of smallpox), which killed or disabled scores of children because of adverse reactions to the vaccine.

One microbiologist following in Ishii's footsteps killed four people and injured hundreds by poisoning hospital food with typhoid fever and dysentery in order to obtain clinical samples for a doctoral thesis. In the 1980s, Ishii's men lied about the safety of HIV-contaminated blood products and killed more than 500 hemophiliacs. Others later experimented on hospitalized children with dubious influenza vaccines. By 1997 scandals caused by former members of Unit 731 or scientists trained by this black fraternity had so defiled the reputation of JNIH that it changed its name to the National Institute of Infectious Diseases.

In his haunting history of Unit 731, Sheldon Harris concludes that "the tradition of indifference to suffering pioneered by the Unit" lives on. It may never be erased. Harris concluded that the medical profession had been irrevocably tainted by anthrax.

In the United States, experiments with anthrax have left an equally troubling legacy: the mandated use of anthrax vaccine for U.S. troops. Since 1987 the U.S. government has stockpiled anthrax vaccine for the simple reason that it still regards anthrax as the number one "threat agent." But the vaccine in question, which was first licensed in 1971, has never been up to the task. Nor was it ever licensed to combat aerosol exposure (the very kind of anthrax that would be delivered in a biological attack). In fact, the vaccine only clearly works against cutaneous anthrax and is largely recommended for veterinarians and woolsorters. Nevertheless the government started a mandatory mass vaccination campaign for the armed forces in 1997. In the process, government scientists, the Food and Drug Administration (FDA), and the Centers for Disease Control blithely ignored the absence of published studies proving the vaccine's efficacy and safety. Like Ishii, they just experimented.

As soon as the shots began (a series of six injections over 18 months plus yearly boosters), thousands of GIs started to report a series of consistent reactions: fatigue, muscle and joint pains, and cognitive impairment. According to military statistics, as many as 50 percent of the vaccinated experienced pain and a rash at the injection site. Within short order, half a dozen healthy soldiers died from a rare pneumonia associated with immune suppression. Subsequent studies have found an oil called squalene in the vaccine that may be linked to Gulf War syndrome. Of the more than 1.3 million military personnel who have been forced to get "Vaccine A," more than 3,000 have recorded serious reactions. In 2002 the General Accounting Office found that 69 percent of the trained and experienced pilots and aircrew members in the National Guard and Reserve were rethinking their military careers because of the vaccine's notoriety. Anthrax had become Osama Bin

Laden's fifth column. In his well-documented 2004 book, *Vaccine A,* the American journalist Gary Matsumoto concludes that military doctors in both England and the United States did not tell soldiers "they were getting an unlicensed immunization, let alone one that contains a substance shown in peer-reviewed scientific literature to be capable of causing incurable if not fatal disease."

In recent years Vaccine A and its defenders created more illness and disrupted more lives than the Amerithrax attack. Whenever soldiers objected and raised the issue of informed consent, the brass became punitive. Of the nearly 600 people who refused the vaccine, about 400 received nonjudicial discipline or were docked pay. Others got less than honorable discharges. At least 51 soldiers were court-martialed, including Captain John Buck, a military physician. When Buck refused to give shots on the grounds that the vaccine had not been properly tested, his superiors put him on trial in 2001. The military judge demoted Buck and fined the officer $21,000. "I was at the crossroads between the oath of an officer and the oath of a physician," said Buck during his trial. "The only way I could have peace about the apparent conflict was to do what I knew to be right as a physician and to stare down the barrel of the gun with the courage of an officer."

The maker of the vaccine, Bioport, proved as unreliable as its product. It repeatedly failed to demonstrate sound manufacturing practices, and as a consequence the government quarantined more than 6 million doses. When the FDA ordered the company to update its label on adverse effects in 2002, the percentage of possible bad reactions rose from 0.2 percent to between 5 and 35 percent—an incredible increase of 17,400 percent. During the Amerithrax scare, the government offered the vaccine to 10,000 people who were also taking antibiotics. Not surprisingly, only 130 elected to take the controversial vaccine.

In 2003 a federal judge ended the friendly fire from anthrax by ruling that "absent consent or presidential waiver, the United States cannot demand that members of the armed forces also serve as guinea pigs for experimental drugs." Although the government got the order quashed, the judge soldiered on (as did a lawsuit) and came back with an identical ruling in 2004: The vaccine wasn't safe and no soldier should have to take it against his or her will. The U.S. government now proposes to buy 75 million doses of another untested anthrax vaccine.

Meryl Nass, a physician who takes her Hippocratic oath seriously, challenges the reasoning behind these gross expenditures ($1 billion). She believes a new anthrax vaccine (as currently designed) won't save lives and surely won't deter terrorists. Now that science has designed germs that are vaccine resistant as well as special chimeras that can cause more than one disease (such as the Legionnaires' disease–multiple sclerosis bacterium), stockpiles of Vaccine A don't make a lot of sense.

Nass also knows the counterargument. Defenders of the biodefense establishment say more vaccine is needed in order to save an urban population that may be attacked by anthrax. But as Nass notes, no vaccine is 100 percent effective. Immunity for anthrax lasts less than a year and then decreases over time. An anthrax spore can live for more than 100 years, and there are strains that can defy every vaccine and drug available. In real time, many people and many animals would die in any serious anthrax attack. In simple terms, "Vaccination will not solve the problem of anthrax contamination."

Nor will more high-security laboratories do the trick. After Amerithrax the U.S. government began a $2.5-billion building boom for bioterrorism defense laboratories, with the idea of having more scientists studying the threat posed by anthrax and custom-made germs. If all 20 proposed labs are built, another 6,000 scientists and technicians will have access to some of the world's most dangerous pathogens.

Richard Ebright, a microbiologist at the Waksman Institute at Rutgers University in Piscataway, New Jersey, believes the labs are a hazardous overreaction. He argues that more Level 4 (top-security) labs will not only become a source of deliberate or accidental leaks (even the

SARS virus escaped from three different Level 4 labs in Asia) but will also produce more individuals comfortable with anthrax's dark legacy. Ebright pointedly adds that a 2001 study found that professional researchers were responsible for most of the 21 identified germ attacks in this century. "The substantial majority were research or medical personnel which is what you would suspect," he told *The New York Times* in 2004. At least half a dozen other prominent scientists agree with Ebright: "We find the idea of a government-sponsored, large-scale multi-site building boom frightening."

So, too, are the latest techniques for creating designer pathogens. They are the genetic engineering of organisms and synthetic biology, which is the science of building organisms from component parts, Lego-style. Thanks to recent genetic advances, scientists can now make a flu virus transmissible from human to human. They can build a mouse pox virus that even slays mice drugged with antivirals or fortified by vaccines. They can construct a polio virus from scratch. They have engineered a more deadly rabbit pox virus. In 1992 they even beefed up a common strain of soil bacterium, *Klebsiella planticora,* to help decompose plant stubble and compost. But the engineered strain had one fatal problem: It killed all plant growth. Scientists can also make a stealth virus that, the CIA says, "could lie dormant inside the victim for an extended period before being triggered." In fact, the CIA now claims that technology has made so many man-made invaders possible that "it is no longer possible to distinguish between legitimate biological research activities and production of advanced Biological Warfare agents."

The economics of biological warfare are almost as totalitarian as anthrax's corruption of the scientific community. A 1996 report by NATO found that anthrax held a definite cost advantage over conventional and nuclear weapons. It simply produced a higher casualty rate at less cost than any other weapon of mass destruction. In fact, a small airplane dispersing 100 kilograms (220 pounds) of anthrax spores over the city of Washington, D.C., would be more lethal than a hydrogen bomb. On a clear night such an attack could kill 1 million to 3 million people in an area of 780 square kilometers (300 square miles) and leave property intact. In contrast a hydrogen bomb would wipe out only a million or so folks and flatten all useful infrastructure.

In 1997 the U.S. Centres for Disease Control also gave anthrax the economic thumbs-up. The CDC study compared the effectiveness of three man-made invaders as bioweapons: tularemia (a debilitating illness with a 30 percent mortality), brucellosis (a nasty bacterium that causes blood poisoning), and good old anthrax. Brucellosis and tularemia would kill thousands and inflict heavy economic losses (nearly $500 million), but nothing would prove as devastating as anthrax. The CDC estimated that for each 100,000 persons infected, an anthrax invasion would cost the economy $26.2 billion, equivalent to the total economic sabotage unleashed by Britain's BSE and FMD outbreaks combined.

In a submission on bioterrorism to the U.S. Congress in 2001, Nass spelled out an obvious truth: "Our species could be obliterated from the face of the earth using technologies widely available today." She then quoted the Nobel laureate Joshua Lederberg, who recently observed that "there is no technical solution to the problem of biological weapons." It is a problem that needs a moral and ethical solution, says Lederberg.

But the history of anthrax shows that moral and ethical considerations don't mean much when science is pursued for science's sake under a veil of government secrecy. In the brave new world of "deliberately emerging pathogens," ordinary citizens might well ask the same question posed by the novelist Aldous Huxley in 1932: "What's the point of truth or beauty or knowledge when anthrax bombs are popping all around you?" Huxley's answer was a world of unending tyrannies.

SEVEN

Marine Invaders: Cholera's Children

"While water quantity has been the major issue of the 20th century, water quality has been neglected to the point of catastrophe."
—Aaron T. Wolf

Salomon Medina knows that there is no love in the time of cholera. In 1992 the 60-year-old Warao elder encountered a Bengali invader on the shores of the Mariusa River in Venezuela that quickly unsettled his aboriginal community with explosive diarrhea. Within short order the microbial immigrant dispatched two elders, including the chief medicine man. "We were eating well, eating well and then all of a sudden we started shitting all the time," explained Medina. "We were living as we always had. Look, we were happy. And even though we were fine, our friend Santiago Rivera started shitting in the middle of the night. He shitted four times. He was getting really sick. 'I'm getting really weak, I'm getting really weak.' When he grew silent after saying those words, he died."

Cholera, an Asian globe-trotter and the world's most feared water-borne infection, had not appeared in Latin America for nearly 70 years.

But globalization has a knack for waking the dead. In the early 1990s, cholera made a rousing comeback, eventually killing 10,000 poor people across South America. The bacterium's invisible toxins immobilize the gut so systematically that all precious bodily fluids sweep through the body like water through a ruptured sewer pipe.

Medina, like 19th-century slum dwellers before him in London and Moscow, clearly felt the result: "We were shitting, the guy was shitting, shitting, and when he shitted again he passed out. 'I'm going'—those were his last words. Look, then horrible cramps would shake our bodies and people would die right away, that was how they died. Another, another, and another died, and when dawn came another died and another was shitting." All in all, 500 aboriginals left this world with nary an obituary. Cholera, easily the world's foremost marine invader, had just done what invaders do best: remove more diversity.

Cholera may be the world's most celebrated marine invader but it is certainly not the only one sailing the world's waters. In fact, its fungal, protozoal, and viral relatives are now splashing up a storm in fresh and salty pools around the globe. In their wake, they are leaving massive algal blooms, devastating marine plagues, dead coral reefs, and rearranged worlds that make mere chaos look good. Fish farming, ship traffic, the aquarium trade, mass migration, and much human defecation upstream have all contributed to this unparalleled fouling, as has climate change. Unknown to most landlubbers, the ocean is the world's largest carbon sink. But it has now absorbed so much carbon dioxide from our fossil fuel addiction that its waters are becoming both warmer and more acidic. Soon the shells of marine creatures will soften, dissolve, and become targets for traveling pathogens at unfathomable rates. Like it or not, our oceans and lakes have become as disease-ridden as our hospitals, as unhealthy as our livestock, and as vulnerable as our crops.

To appreciate how topsy-turvy the world's waters have become, take a quick visit to San Francisco Bay. The estuary may still look hip and smart, but under the surface, it's a global jungle. Every shallow habitat has been hijacked or occupied by more than 220 alien species that arrived on cargo ships. Like European imperialists, they have exterminated or neutralized local fish and plants. The bay boasts zooplankton from Japan, jellyfish from Europe, cord grass from the Atlantic, mitten crabs from China, pill bugs from New Zealand, bait worms from Maine, soft crab from Chesapeake Bay, clawed frogs from South Africa—and every year another four migrants arrive and add to the shambles. Nobody has counted, but each and every invader brought along a bucket of bacteria, viruses, and parasites that now lurk below the waters. San Francisco Bay, like the waters around Hong Kong or Miami, has become a hellish mishmash of creatures and a nightmarish glimpse of things to come.

We are doing to water what we do to some happy Disney-like aquarium: We mix and match creatures, simplify environments, sprinkle lots of nutrients, and then are surprised when the whole kit and caboodle dies and floats to the top or explodes into a smelly algal bloom. Aquarium owners routinely flush their failed experiments down the drain. But no such option exists for the world's oceans and lakes.

The ocean occupies 70 percent of the planet's surface. More than 2 billion people have an ocean view, and 14 of the world's largest cities sit on or near the ocean. Despite the dangerously looted state of the world's fisheries, we still depend on the ocean for the majority of our protein. Any way you look at it, the ocean matters.

Yet these vast salt waters, from which all life and biodiversity emerged, aren't getting the respect an elder deserves. Each year people, cities, and corporations dump vast volumes of wastewater into ocean and coastal waterways. The United States, a relatively clean country, flushes 11 billion liters (3 billion gallons) of animal waste, pesticides, nitrogen, drugs, and other refuse into the ocean every year. As a consequence the Gulf of Mexico supports a "dead zone" the size of New Jersey where nothing grows. Since the 1970s, agricultural waste has created more than 150 oxygen-stealing dead zones around the world.

In a 2002 assessment of the future of ocean life, arguably the largest and most resilient ecosystem on earth, Peter Verity, a researcher at the Skidaway Institute of Oceanography in Savannah, Georgia, found nothing but man-made trouble. Unprecedented marine plagues, algal blooms, relentless invasions, and rising pollution headed the list, and there was more trouble to come. "There is a flock of miner's canaries signaling change, degradation and cryptic reorganization of marine communities," he warned. Among biologists, Verity is probably an optimist.

Freshwater sources aren't faring much better. Rita Colwell, the former director of the U.S. National Science Foundation and a celebrated cholera biologist, has repeatedly spelled out the grim math. Only 2.5 percent of the planet's water is fresh, and one-third of that is frozen in ice caps or glaciers (although climate change may be quickly releasing it). Another one-third comes in the form of monsoons or floods and can't really be captured. That leaves the remaining third, less than eight-tenths of a percent of the total water on earth, for human consumption. We know that 1 billion people lack decent drinking water and that another 3 billion don't have adequate sanitation. We may add another 2 billion to 5 billion water drinkers to the planet by 2050. As a consequence, the world already boasts more water refugees than it does war refugees. Yet, as Colwell notes, "the cycling of water and our global interconnections mean that all of us are living downstream."

No invader illustrates the global plight of living downstream better than cholera. It's the world's first truly global disease and seems to be getting stronger. Since the 19th century it has buried millions, generated seven pandemics, filled libraries with somber treatises on bad water, championed the plumbing trade, and generally exposed poor watershed management everywhere. Cholera, in short, has been a reliable barometer of how we invade, transport, and pollute water. Thanks to our ambitions of trade and empire, it's more geographically widespread and persistent than ever before.

The bacterium *Vibrio cholerae* is a seasonal and very complex serial killer that started out as a native of the lower Ganges River. The invader, which comes in three distinct guises and can even shrink to less than 1 percent of its normal size in cold water, flourishes in dirty, nutrient-rich pools. Its favorite resort remains the brackish coastal estuaries off Bangladesh, where it attaches itself to tiny shrimplike creatures called copepods. These ubiquitous crustaceans, not much bigger in diameter than the width of a sewing thread, eat algae (the base of all marine food chains) and help transport cholera from one ocean neighborhood to another. In exchange for the ride, the bacterium helps to rupture egg casings for female copepods.

In humans the cholera germ can set off a diabolical elimination cycle. Once in the gut, the bacterium releases a toxin that forces the body to give up its water and precious salts in one evacuation after another. In half a day, cholera can drain 10 percent of the water from a person's body. Although the infection can be treated economically with a rehydrating solution of salt and sugar (a remedy first deployed only 30 years ago), the poor are still left to fend for themselves. The promiscuous use of antibiotics for cholera has made many strains of the disease drug resistant. For reasons nobody truly understands, cholera particularly affects people with Type O blood types, which explains why cholera ravaged Europeans in the 19th century and Venezuelan aboriginal people in the last decade.

It takes a special brew of bad water to get cholera cooking. The disease generally runs amok whenever copepod populations explode, mostly during algal blooms. When warm temperatures grow fantastic pools of marine plankton from August through October, folks in Bengal start running to the outhouse or the cholera hospital. In the Bangladeshi town of Matlab, near the Bay of Bengal, medical and weather records show that outbreaks of cholera invariably follow drought or monsoons: Dry spells concentrate the organism in stagnant pools of water, and extreme rains contaminate drinking water with the bacterium. (Floodwaters generally carry 1,000 times more waterborne diseases than water sources under normal conditions.)

Cholera knows how to take advantage of extreme commerce and extreme migrations as well as extreme weather. For several thousand years it was largely confined to the vast river deltas of the Indian subcontinent, where even a Hindu goddess celebrates its presence. It made the odd foray to Mecca or China on trading ships but remained essentially a Bengali resident until the 19th century. With the help of the British Empire, cholera started its grand global tour after a religious festival on the banks of the Ganges in 1817. Returning pilgrims spread the bad news throughout India, and the British army alone lost 10,000 troops. Along trade routes to Iran and Russia, Indian merchants defecated to death and unknowingly contaminated more water. Kegs of fouled water on trade ships spread the plague from port to port.

Cholera, never a slacker, hitchhiked on the empire's troop movements, aboard its tea and opium ships, down irrigation canals, and along India's new national railway system. By 1832 it was a global bane. The German poet Heinrich Heine even met cholera at a Paris ball: "Suddenly the gayest of the harlequins collapsed, cold in limbs, and underneath his mask, violet blue in the face.... Soon the public halls were filled with dead bodies, sewed in sacks for want of coffins."

In Europe's fetid slums and sewers, cholera found foul water as hospitable as the feces-rich Ganges. As the disease swept away the poor, the medical establishment assured the public that cholera couldn't be contagious, partly because industrial magnates couldn't entertain the thought of trade-stopping quarantines. The British *Annual Register* noted that "the cholera left medical men as it had found them—confirmed in the most opposite opinions or in total ignorance as to its nature, its cure and cause." After a quarter of a million people died in Russia, the poor became so enraged at government indifference (they correctly saw the failure to combat cholera as a plot to kill them) that some formed a revolutionary club called the Cholerics and waged war against the czar. Biological terror, sooner or later, begets political terror.

Since its 1817 global debut, cholera has traveled around the world in six more pandemics. Each time the invader has broadened its geographical scope to kill or sicken more people, and each time it has traveled more quickly. In the 1830s, it took three months for cholera to hop from Hamburg, Germany, to northern England. But in 1848, steamships and sickly immigrants give cholera first-class passage from Poland to New Orleans in just seven weeks. By 2002 contaminated airline food was allowing passengers to move cholera about the globe in just hours.

By the 1850s, cholera had made quite a name for itself. The disease had killed millions and highlighted the terrible state of water sanitation. The father of epidemiology, John Snow, defiantly removed the handle from a contaminated water well on Broadstreet in London; the resulting drop in infections proved that cholera was, after all, a waterborne disease. Snow also observed another powerful truth: "Epidemics of cholera follow major routes of commerce. The disease always appears first at the seaports when extending into islands or continents."

But as European imperialism cemented global networks, cholera just got stronger, gaining access to human stomachs in Latin America, North America, Australia, Europe, and Asia. The seventh pandemic, which began in 1961, best shows how the speed of modern life can advance an adaptable marine invader.

After a global lull of nearly 38 years, cholera appeared in new guise in Indonesia's slums and then waltzed through Asia in the 1960s. The new strain, called El Tor, had first appeared in a quarantine camp in the Sinai peninsula in 1901. It resisted several antibiotics, defied chlorine, lasted longer in the environment, and produced a higher number of asymptomatic carriers than earlier strains. That meant more people could unknowingly contaminate more food and water with the bacterium than ever before. By the 1970s, cholera had entered Russia and, for the first time, West Africa. Mass migrations of African war refugees soon took cholera to almost every nation on that continent. By the 1980s, cholera had besieged the poor in a record 36 countries around the globe.

In 1991 the pandemic jumped continents. That's when a Chinese freighter dumped its ballast water off the coast of Peru and with it a stowaway cargo of cholera. (Research by the cholera expert Rita Colwell later found that every liter or quart of ballast water contains 1 billion bacteria and 7 billion viruslike agents, but customs officers never ask for their ID.) From coastal waters and contaminated seafood, the disease ripped through the slums of Lima and hit seven other countries, disrupting several billion dollars' worth of tourism, trade, and travel. Aided by poor water systems, foul surface water, and even vegetables washed with unclean water, cholera eventually infected 1.4 million Latin Americans, killing more than 10,000 people.

Everywhere cholera traveled during the seventh pandemic, the invader illuminated a host of global vulnerabilities as well as the pathways for future invaders. In 10 crowded Ukrainian cities, it exposed poor health services and poor hygiene, prompting one politician to declare that "the spread of cholera and other infectious disease is the calling card of an economy in trouble." In the relief camps of Rwandan refugees, it played on despair and contaminated water to bury 12,000 people. On a flight from Los Angeles to Buenos Aires, 31 passengers encountered a nonpaying passenger: cholera. Thanks to the industry of the merchant marine and its careless ballast habits, almost every major coastal estuary from Chesapeake Bay to the China Sea now supports populations of *Vibrio cholerae.*

A world that builds bigger slums, degrades public health systems, pollutes coastal waters, empties ballast water in ports, throws half a billion travelers across international borders, and sends 100 million people on various economic and political errands every year is really a world committed to cholera's prosperity and the glorious growth of all waterborne invaders. Not surprisingly, the seventh pandemic has lasted longer than any previous one (45 years so far) and has affected more nations than all of cholera's earlier visitations. Although doctors have arguably limited the deadliness of cholera outbreaks with simple salt and sugar therapies or by advising Bengali women to use folded sari cloth to filter out cholera transporters such as copepods, the triumphant germ now has achieved a greater global penetration than ever before. Its Latin American visit left a bill of $200 billion for water cleaning.

Two British public health experts, Kelley Lee and Richard Dodgson, concluded in 2000 that cholera "is a mirror for understanding the nature of globalization" and that it is poised to teach us more hard lessons. Science suggests they are right. In 1992 a new strain (*Vibrio cholerae* 0139 Bengal) emerged on the banks of the Ganges, where 29 cities now empty 325 million liters (345 million gallons) of raw waste a day. Cholera's latest strain is hardier than El Tor. Like its forebears, it likes to travel in ballast water, contaminated fish, and migrating humans searching for clean water.

Cholera's children have been just as busy unsettling ocean life too. Although little is known about marine pathogens, Drew Harvell, an evolutionary ecologist at Cornell University, has been documenting the rising carnage among corals, sea urchins, mollusks, and other marine creatures with increasing alarm. "Many of these epidemics are like lightning strikes: They hit quickly, unexpectedly, and run through a population and then are gone."

In fact, marine invaders don't behave like land-based invading pathogens. Water invaders, which account for 80 percent of all infectious diseases, can outrun most of their sluggish land-based cousins. A viral fish eater can travel up to 10,000 kilometers (6,200 miles) a year; one type of canine distemper, for example, covers 3,000 kilometers (1,860 miles) a year as it attacks seals. (The West Nile virus can traverse only 1,000 kilometers [620 miles] a year on land, even with the help of birds and mosquitoes.) Marine diseases also have a vast open expanse to swim in. No mountains or forests stand in the way to slow the spread of microbial pollution. And none of the standard disease-fighting tools adapted for terrestrial warfare actually work in an ocean. "Vaccines, quarantines, and culls can't be applied in the oceans," notes Harvell. A sick ocean, in other words, has nowhere to run.

Accounts of the epidemics now killing marine life in record numbers read like scenes from florid science fiction. In 1995 a viral member of the huge herpes family worked its way through millions of sardines off the coast of Western Australia, much like a bushfire. In the ocean the virus moved at incredible speeds of 30 kilometers (18 miles) a day, eventually infecting a stretch of coastline as long as California's. The viral invader probably hitched a ride on imported protein fed to factory-farmed tuna. It then took off into the wild.

Seals are dying just as fast as sardines. In 1980 a distemper virus ambushed 20,000 harbor seals, nearly 60 percent of the population in northern Europe. It filled their lungs with fluid, planted ulcers on their skin, and burned them up with fever. Laden with PCBs and other industrial treats, the seals were as immunosuppressed as AIDS patients. Given such a burden, the mammals couldn't mount much of a fight against the invader. Peter Ross, a Canadian biologist who has studied the tortured connection between successful viruses and failing immune systems, calls the harbor seals another proverbial canary in the great marine mine. (One wonders just how many sick canaries this planet can support.) He emphasizes that the same waters that killed the seals also put culinary delights on European tables. Ross sees the diseased seals as another sorry window into the "state of the world's oceans and the commercial fisheries."

Green turtles have also joined the club of reluctant sentinels. They've swum the earth's oceans for 200 million years, and many cultures consider them a totem of wisdom. But another herpes invader called fibropapillomatosis threatens to put this endangered species all the more at risk. The virus, which is probably spread in nesting areas, transforms the face and neck of the turtles into a mass of fist-sized tumors and warts. Unable to see or eat, the turtles waste away within a year. Many of the reptiles actually bear an uncanny resemblance to AIDS patients with Kaposi's sarcoma, another herpes hit man that blotches the skin with cancerous, leprosylike spots.

In the last 20 years, this viral outbreak has sunk turtle populations in Hawaii, the Caribbean, and Florida. Now it's spreading to other turtle species. Most researchers agree that the incidence of deformed and dying turtles appears to be greatest in polluted waters. They now suspect that the invader is an old viral dog that is playing lethal new tricks among reptiles whose immunity has been lowered by dirty water.

But that's just the beginning of the new marine plagues. A shell disease is now disfiguring and weakening the tasty American lobster in the Gulf of Maine. Warmer water, thanks to climate change, adds up to more physiological stress, which in turn leads to more disease vulnerability. (Lobster fans should know that *Homarus americanus* specimens blotted with the bacterial infection are not wasted; they are turned into lobster salad or lobster Newburg.) Dumbfounded scientists and fishermen still don't know what to do about the epidemic and have even taken the crustaceans to veterinarians for help. "We could not have predicted it a decade ago," lamented one biologist to *The New York Times.*

Off the coast of California, sea otters are being driven mad by *Toxoplasma gondii,* a one-celled parasite commonly shed in cat, possum, and livestock excrement. Toxoplasmosis is a fairly common infection spread by water fouled by oocysts. Although the disease rarely gives cats more than a fever, it can cause big trouble in humans with weak immune systems. The parasite can also cause birth defects or force an unwanted abortion. It may even be responsible for a number of cryptic neurological illnesses. Toxoplasmosis inflames the brains of AIDS patients, producing hallucinations and delusions.

The protozoan seems to have the same effect on otter nerve tissue. "The animals are often found alive and are suffering seizures, showing obvious signs of damage to their brains. They can't hold food and can't take care of themselves, and their eyes are dilated," reported David Jessup, a senior California wildlife veterinarian, in 2003.

Between 1998 and 2001 state biologists collected 105 sea otter carcasses off beaches near Monterey Bay and autopsied them. Most of the otters had been in their prime; swollen brains caused by *T. gondii* infection appeared to be the leading cause of death. The researchers offered a number of damning explanations. Some thought feral cat and possum populations near factory dairy farms were to blame. Others singled out pet lovers flushing infected kitty litter down the toilet. A steady decline in natural water filtration thanks to insatiable human consumption may have also thrown a protozoan pathogen "not likely to have been abundant in the marine environment centuries ago" into the otters' coastal bathtub. In any case, the abundance of the parasite probably explains why California's otter population (about 2,300 animals) has declined or remained static for the last decade. *T. gondii* might also explain the catastrophic demise of 80 percent of Steller sea lions and northern fur seals in the north Pacific.

Some water bodies are more blighted than others. The Caribbean, a global symbol of fun and sun, has witnessed a succession of 30 different marine plagues in the last 30 years. First a mysterious pathogen killed off the region's most important algae eater, the long-spined sea urchin. The prickly herbivores lost their spines, turned black, and sat listless in the open during the day. From 1983 to 1984, nearly 90 percent of the population died out. Their numbers still haven't recovered.

Next came a plague that killed off sea grasses, followed by a variety of potent coral killers. White-band disease and black-band disease polished off coral formations as much as 7,000 years old. Staghorn coral, once the dominant species on many Caribbean reefs, has been nearly wiped out by one epidemic after another. One of the most efficient killers has been a soilborne fungus, *Aspergillus sydowii*, a smotherer of the immunocompromised. (A closely related species, *Aspergillus fumigatus*, preys on AIDS patients and young cancer patients, and its incidence too is rising in hospitals.) The fungus may have arrived on dust swept from the African continent in the 1990s (along with live locusts, 110 types of unidentified bacteria, and other stuff) and has since polished off 90 percent of soft coral communities in the Caribbean and Florida Keys.

The coral die-off in the Caribbean signaled a global sickening. A raft of bacterial and fungal diseases are now denuding coral reefs along the coasts of India, Florida, and Indonesia. Modest increases in the concentration of human and animal excrement and agricultural waste in local waters all seem to exacerbate disease severity and die-off rates. The impacts of these serial plagues, which often overwhelm corals with toxins, have garnered little fanfare. Corals grow slowly, live for centuries, and reproduce when they feel like it. Although they occupy only about 1 percent of the planet's surface, they claim a biological diversity greater than the rain forests'. In addition to their iridescent beauty, they protect seashores and support a third of the ocean's fish species. Now hammered by pollution, overfishing, warming temperatures, and as many as 15 different plagues, the world's ocean reefs are fast losing their diversity, which remains nature's most cost-effective vaccine against disease.

The rising tide of marine diseases has been accompanied by startling blooms of toxic algae. Red tides, called HABs (harmful algal blooms) by marine scientists, used to be rare occurrences that required the odd Biblical notation. But now they are popping up everywhere in the ocean, killing fish and turning mollusks into toxic bombs. Algae, 3-billion-year-old denizens of the planet with 50,000 cousins, make 50 percent of the world's oxygen. But when fed too many nutrients, they can bloom crazily into red, green, and brown tides and go on vicious killing sprees. As many as 50 species can produce deadly surprises, such as paralytic, amnesiac, or neurotoxic shellfish poisoning. Don Anderson, who researches algal species at the Woods Hole Oceanographic Institute in Massachusetts, knows that little things can add up to big messes when they explode and invade new environments: "There are more toxic algal species, more algal toxins, more fisheries resources affected, more food-web disruption and more economic losses from harmful algal blooms than ever before."

China, home to the world's most intense pollution of coastal waters, is a case in point. Between 1952 and 1998, more than 320 HABs swelled offshore, killing millions of jellyfish, shrimp, scallops, clams, and other marine organisms and causing hundreds of millions of dollars' worth of damage. Since 1998 some blooms have occupied areas larger than Taiwan.

Chinese HAB watchers blame much of the ocean killing on "increased urbanization." In fact, the polluted waters off Hong Kong and Guangdong province are hot algal zones. Sea creatures aren't the only casualties. In the last 40 years, toxic shellfish have poisoned 1,800 seafood lovers (a very conservative Chinese tally) and killed at least 30 people. Tian Yan, a researcher at the Chinese Institute of Oceanography, offered this uncharacteristically truthful assessment: "This increase in HAB frequency, scale, and economic loss in China is drawing great attention from the government and the public." No one has seriously drawn up a plan to limit the sewage and other waste poured into China's waters, but the study of HABs has become a trendy science.

The gloomy blooms are also provoking other hidden menaces. Waste from North Carolina hog farms flushed into the ocean stirred up the cell from hell in the 1990s. Dinoflagellates are resilient algae eaters that come in at least 85 toxic species. The most notorious is *Pfiesteria piscidia* (literally "fish killer"), which comes in 24 different guises ranging from a hibernating cyst to a two-tailed flagellate that stuns its prey with neurotoxins. It can masquerade as a plant or lie dormant for years. It can exist in fresh or salt water. It mostly feeds on algae, but whenever algal blooms explode, toxic dinos go on a feeding frenzy too, killing millions of fish by sucking away their flesh. Since the 1990s, *Pfiesteria* has killed billions of fish from Florida to Delaware and poisoned thousands of shellfish eaters. Because its airborne toxins can produce pain, memory loss, immune system failure, and personality changes, *Pfiesteria* has even earned a Biohazard Level 3 status. (The highest level is 4, where Ebola ranks.) Fishermen who have sailed through toxic dino blooms have passed out and later watched their health unravel. Toxic dinos are now on the move and have invaded waters off Scandinavia, Latvia, and New Zealand.

The progressive sickening of coral reefs and marine mammals coupled with poisonous algal blooms remains for most people an obscure tidbit of information. But when a cat parasite invades a sea otter's brain and an AIDS-like virus turns turtles into sea lepers, when lobsters start acquiring staphlike infections and an opportunistic fungus wipes out coral reefs, immunocompromised kids, and cancer patients, nature is probably trying to get our attention. Drew Harvell now suspects that we have set in motion a hurricane of marine misfortunes by "favoring the dispersal and development of new variants of wildlife pathogens through global commerce, climate shifts, and changes in host density and habitat quality."

We probably owe the disastrous spread of a lot of these infectious marine invaders to ballast water. Every year 45,000 cargo ships transport 10 billion tonnes (11 billion tons) of ballast water around the globe. Drawn from ports and bays, this water contains a veritable zoo of material: fish, mollusks, worms, algae, cholera, and all sorts of other microorganisms. These cargo ships are simply "floating biological islands." James Carlton, one of the world's foremost experts on ocean invaders, calculates that "on any given day, there may be as many as 7,000 species in motion around the world via tens of thousands of ships."

Given that the number of commercial ships plying the ocean doubles about every 20 years, the spread of marine diseases and other species can expect a future as positive as oil stocks. Some members of the U.S. Health Resources and Services Administration are now so worried that one even compared ballast water to a "biological time bomb liable to engender significant disease in vulnerable populations anywhere in the world."

Another key distributor of disease has been fish farming, the marine equivalent of the livestock revolution. As fishing stocks have perilously declined in the last decade, factory-style fish production has doubled, and it now serves up nearly 50 percent of the world's animal protein production. The American arm of the industry alone raises and trades more than 100 species. Like the livestock revolution, the blue revolution farms exotic species as well as genetically altered stock and thinks nothing about moving eggs, equipment, and frozen fillets about the world as if biology never mattered. Not surprisingly, fish farmers spend a lot of time fighting disease outbreaks and attending biosecurity forums.

To date aquaculture has attracted more microbial attention and pathogenic invaders than an immunocompromised AIDS patient. This dirty industry overcrowds fish into small ocean-based pens or nets and nourishes them with manufactured feed and antibiotics. The excrement of the penned fish then fouls local water with waste rich in bacteria and viruses. Yet global traders and fish farm promoters still feign surprise when waves of invaders strike their "floating pig farms."

Norway was probably the first to prove the disease-spreading abilities of aquaculture with its salmon farms. Overcrowding spawned the infectious salmon anemia virus. Boats, sea lice, fish feces, and nets then spread the virus to wild salmon as well as to fish farms in Scotland and Canada, where epidemics killed off 90 percent of the farmed fish. The virus is now found in sea trout and rainbow trout. In an attempt to quell the explosion of diseases spreading from salmon farms to wild fish, the Norwegian government literally poisoned scores of rivers and streams in order to reboot them ecologically. It didn't work.

In Japan the carp industry has been battling the highly contagious koi herpes virus, which first popped up in Israel in 1997. The virus targets the gills, stops the fish from eating, and then disorients it. Die-offs average 90 percent. It can be spread by mud, buckets, plants, and water. Fish farms in Taiwan, Indonesia, the Netherlands, and the United States are now battling the invader.

The tiger shrimp industry has become a biological war zone as factories in Ecuador, China, Mexico, and India exchange one viral troublemaker after another. China, which employs nearly a million people in shrimp farming, lost more than 60 percent of its production to epidemics in the 1990s. "The diseases spread so quickly that if a few shrimp were found dead or diseased, then a few days later the whole pond would be dead or dying," reported one chronicler. Most of China's shrimp industry lies downstream of 43 industrial cities that spew 6 billion tonnes (6.6 billion tons) of waste and sewage into coastal waters every year. "Unfortunately many shrimp ponds are concentrated near estuaries where pollution waste drains." The impact of this viral explosion on wild and native crustaceans is anyone's guess, but scientists aren't predicting any boon for biodiversity.

At a 2001 conference held by the colorful World Association for the History of Veterinary Medicine, the Norwegian fish doctor Tore Hastein asked what history had taught mankind about the spread of aquatic animal diseases. His blunt answer was, "Probably nothing." After telling how the live trade in fish and fish eggs had spawned an epidemic of spring viremia in carp in the 18th century and then brought a crayfish plague from the United States to Europe, he concluded that the incalculable number of marine diseases ignited by careless trade in fish, oysters, and shrimp had been a grand lesson for the deaf and blind: "The shrimp industry seems to be making the same mistakes ... in regards to the spread of disease by trade." In fact, annual disease losses among Asian fish farms alone now total $3 billion a year. Some shrimp farmers in American coastal areas have moved their operations to the deserts of Arizona in hopes of finding a respite from viral invaders.

The grim literature on marine plagues set off by farms or the aquarium trade is growing by leaps and bounds. (The aquarium business imports 150 million exotic freshwater and marine fish of more than 2,000 species every year to the United States alone.) In the 1950s, the restaurant seafood trade unwittingly dumped a deadly protozoan into

Chesapeake Bay that finished off the American oyster. The Asian gudgeon, a baitfish introduced to Europe in the 1960s, carried parasites that massacred native minnows and now threaten salmon. The abalone trade carelessly introduced a parasitic worm from South Africa into coastal waters off California, where it's now deforming the locals. We are reshuffling every marine organism on the planet but don't have a clue how the new arrangements will work or if they will even be hospitable.

Meanwhile the biological storm continues to churn up trouble for humans too. Consider the silent progress of *Cryptosporidium parvum*. Once deemed an exotic parasite, it is now a global resident of fresh and salt water, causing public health scares, deaths among immunocompromised people, and diarrheal diseases in others. It can live in chlorinated water, outwits standard water filtration, and is considered a threat to the U.S. water supply. Following in cholera's footsteps, the parasite shows how pervasive biological invaders have become in water.

OYSTER DECLINES IN CHESAPEAKE BAY

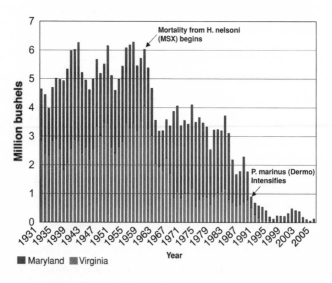

Two invaders have decimated native oyster populations.

Source: *Change Futures Health, Ecological and Economic Dimensions Report.* A Project of The Center for Health and the Global Environment, Harvard Medical School. Sponsored by Swiss Re and United Nations Development Programme. Reprinted by permission.

Crypto, a cousin to malaria, belongs to the order Coccidia, which boasts a multitudinous following of protozoans. An American scientist first discovered the bug in the gut of a mouse in 1907. It later came to the attention of veterinarians as a formidable diarrhea-maker among birds and mammals. When shed in animal stools, crypto takes the form of an oocsyt, a ball with a hard shell that allows the parasite to survive up to six months in pools, rivers, or fresh vegetables. Once ingested, the oocyst breaks apart and starts to reproduce asexually, making many more oocysts. Crypto can turn any gut into a crypto factory. It takes only 10 to 30 oocysts to cause an infection, and a sick calf can make millions in a week.

Until the 1970s, scientists thought that *Cryptosporidium* was found only in dogs, cats, snakes, dingos, and other wildlife. But then a farm child in Tennessee got deathly sick (and slowly recovered). Next a chronically ill farmer came down with explosive watery stools that bore crypto's biological signature. Scientists, a dedicated bunch of linear thinkers, regarded the discovery as a medical oddity. "Few microbiologists envisioned that *Cryptosporidium* would become a major global player in the diarrhea sweepstakes," the celebrated parasite watcher Robert Desowitz wrote later. At the same time, vets started to record startling epidemics of calves sick with diarrhea in burgeoning dairy farms and wondered what the cause was.

In 1982, along came AIDS, or what many Africans still call "slim disease," because of its most inconvenient symptom: relentless diarrhea. To their surprise, American doctors found crypto oocysts in 10 percent of their AIDS patients. Some of the infected were so badly overrun by crypto that they actually had oocysts in their bile ducts and lungs. Doctors couldn't figure out why a parasite commonly found in pigeon feces was shutting down the human gastroinestinal tract. As they later mapped the patterns of anal intercourse among gay men, they got part of their answer: Anal sex can spread invading parasites extremely effectively.

As the protozoans continued to move through factory farms and gay men, they slowly contaminated more and more watercourses. In 1993 a really wet spring illuminated the invader's global progress. Flooding washed out dairy farms upstream from Milwaukee's public water supply (nobody has pinned down the exact source) and sent *Cryptosporidium* into local tap water. The outbreak eventually sent more than 400,000 people squirming to the toilet, emptied local pharmacies of diarrhea remedies (in fact, it was pharmacists who noticed that an epidemic was afoot), hospitalized more than 4,000 people, and killed 100 immuno-compromised individuals. It took a week for health authorities to recognize that a protozoan had besieged the city's bowels.

The invading organism wreaked havoc with the elderly, the sick, and AIDS patients. "I had several requests for assisted suicide and euthanasia. People were just suffering so much," reported Ian Gilson, a Milwaukee physician. Some of the victims lost spleens or kidneys, and transplant patients battled organ rejection.

The epidemic cost the city $96 million in medical bills and lost working days. A new $90-million filtration plant capable of weeding out oocysts was installed. A water quality study later found crypto and giardia (another fierce bowel emptier that causes the misnamed "beaver fever") in almost every watercourse around the city. Another survey concluded that 25,000 water systems serving 92 million people nationwide weren't being operated properly. After the largest outbreak of waterborne disease in the United States, doctors immediately elected crypto to the famous and ever-growing pantheon of "emerging pathogens." Since then crypto has been at the center of hundreds of waterborne outbreaks in North America.

About a third of North America's beef and dairy cows carry crypto, as do more than 40 other animals. Recent scientific studies have even explained the rising tide of crypto epidemics among people: Loads of untreated animal manure produce loads of trouble downstream. Scientists, who often can't call a spade a dirty spade, reported in the *Journal of Water Health* in 2005 that "large microbial loads could be released via heavy precipitation events that produce runoff from livestock manure–applied fields of even modest size." Such events could also "have a significant impact on water bodies within the watershed." (Scientists generally agree that two kinds of crypto have now infiltrated our water supplies: a distinct species carried and spread by human tourists and another one liberally sustained by hordes of industrial livestock.)

Meanwhile the protozoan has acquired global citizenship. Crypto has popped up in both Edinburgh's and Glasgow's drinking water. It has sent thousands to the loo in North Battleford, Saskatchewan. It has been found swimming in the intestines of Egyptian children as well as inside kindergarten students in Anhui province in China. In Ohio's swimming pools, it recently sent 700 to the bathroom. Crypto is now as widespread as cholera.

It is also a ubiquitous marine invader. It can be found in oysters in Chesapeake Bay (a toilet for intensive hog production) as well as in clams on California's coast (a toilet for the dairy industry). Even edible clams in the Adriatic Sea carry crypto. As the *International Journal of Parasitology* put it, clams are now "bioindicators of fecal protozoan contamination." The U.S. Geological Service reported that the hardy protozoan can survive in saltwater at 20 to 30 degrees Celsius (68 to 86 degrees Fahrenheit) for 12 weeks. So people who eat raw or undercooked oysters, clams, mussels, and other shellfish are probably consuming oodles of oocysts for free. And wherever oocysts travel, salmonella, campylobacter, *E. coli* O157 (the kidney buster), and hepatitis A and E as well as the troublesome calciviruses, such as Norwalk, are not far behind.

In fact, the dark voyage of the calcivirus family (four out of five of whose members cause human disease) tells another bleak marine story. This saga started in the early 1970s, when the U.S. navy hired Alvin Smith, an Oregon veterinarian, to study ocean diseases that might give humans trouble. Smith started to look at sea lions because they were then suffering from "an abortion storm with high instances of premature birth." He found the unhappy pinnipeds loaded with the same virus that had caused classic vesicular exanthema of swine (VES). In the 1930s, three separate outbreaks of VES had occurred in California pigs fed raw fish or fish scraps from restaurants. (The disease, much like foot-and-mouth, produces skin blisters.)

Finding a livestock virus in a marine animal troubled Smith, but not agricultural experts. They arrogantly maintained they had eradicated the pig disease in 1956, and they bluntly told Smith that he couldn't call his marine virus VES, even if it looked and behaved like the swine virus. So Smith called the sea lion aborter the San Miguel sea lion virus type 1. When Smith found the same virus in other marine animals, including monk seals, gray whales, and even dolphins, he realized he had stumbled upon a major virus with an ocean reservoir that could colonize land creatures, including California surfers, cattle, and horses.

Some members of the calcivirus family, which look like tiny soccer balls, are celebrated troublemakers. Sapporo and Norwalk, two highly contagious and unpredictable viral invaders of shellfish and humans, are now the globe's number one cause of gastroenteritis and food-borne illness. Norwalk in particular has unsettled thousands in cruise ships, nursing homes, restaurants, and summer camps. In fact, it's not unusual for cruise passengers to experience one or two days of unscheduled retching thanks to the virus. In South Asia and North Africa, the calcivirus hepatitis E, another liver invader, kills 25 percent of the pregnant women it infects. Then there are the animal and marine calciviruses that cause abortion, hepatitis, blistering, and pneumonia. The feline calcivirus, for instance, not only is upsetting the cat world but can also cause infections in dogs and sea lions. Cats, of course, eat lots of fish. Rabbit hemorrhagic disease, a disagreeable bleeder, wiped out most of China's rabbits in the 1980s and went on to dispatch 40 million rabbits in Italy in the course of a few months. It's the only calcivirus not yet associated with human disease, but Smith doesn't expect that exception to hold for long.

Calciviruses, which are some of the smallest life-forms on the planet (a string of 1,400 would be the width of a hair), have all the same characteristics as other cocky biological invaders. After hijacking a cell, any of these RNA viruses can make 10,000 copies of itself in just four to eight hours. One gram (0.035 ounce) of whale feces, for instance, can contain as many as 10 million infectious particles. Infected animals can quickly shower a bay with a calcivirus. (Viruses are the ocean's most abundant organisms.)

Calciviruses are also as flexible as gymnasts. "Virtually every individual calcivirus replicate is a genetic variant from all others and there are more possible viable variants than there are grains of sand in all the deserts and oceans on planet earth," notes Smith. Vaccines are virtually useless against such ever-changing targets. Calciviruses can also survive in frozen environments for years and in 15-degree-Celsius (59-degree-Fahrenheit) water for more than two weeks. In a recent letter to the U.S. Department of Homeland Security, Smith gave calciviruses a glowing review as a biological terror by noting that some calciviruses can be "bulk-grown in the laboratory, easily manipulated genetically, are resistant to normal environmental effects, store easily, have aquatic transmission cycles and reservoirs and very diverse host ranges where a single type can infect 16 species as different as fish and humans."

Smith now suspects that the feeding of raw fish to livestock in the 1930s brought the virus out of the ocean. Shed in livestock manure, the viruses traveled, mutated, and then occupied snakes, frogs, birds, and a host of other species, including humans. (A 2006 study even found them in our blood supply.) The pet food industry probably passed on the virus in the fish it ground up for cats. Contaminated fish meal most likely gave rabbits a hemorrhage-inducing calcivirus that can kill within 24 hours. Smith suspects that waste disposal from factory farming has also done an extraordinarily good job of pouring calciviruses into global waterways.

This biological pollution "has gone essentially unaddressed as has the issue of calciviruses in wastes from food processing including foods from livestock and those from ocean catches and aquaculture operations."

Given its silent global conquest and the virus's ability to cross species, Smith recommends that certain people—women who have miscarried or have given birth to children with birth defects, and patients with hepatitis other than types A through G, diseases that blister hand, foot, and mouth, viral encephalitis, and joint and muscle disease—now ask to be tested for calciviruses.

The sad health of amphibians, the world's oldest and best adapted vertebrates, serves as another bold indicator of water's facility for moving invasive diseases around at a frightening pace. In recent years frog scholars have lamented the silencing of leapers and crawlers in our marshes, swamps, and rivers. Habitat loss, pollution, and logging have all sung roles in this bleak opera, but the leading diva appears to be a nasty fungus with a complex name: *Batrachochytrium dendrobatidis*.

First identified in 1998 by the wildlife biologist Peter Dsazak, the fungus has since been associated with mass die-offs of amphibians in North America, Costa Rica, Panama, and Australia. One group of Australian researchers alarmingly noted, "This is the first wildlife disease to emerge on a global scale that affects an entire class of vertebrates and is associated with mass mortalities, population declines and species extinctions." James Collins, an Arizona ecologist who suspects that the bait trade may be spreading other deadly pathogens, adds a salient point: "Amphibians are clearly vulnerable to environmental change, but the fact that they are being so broadly affected by emerging diseases is telling us that emerging diseases are having a major effect on the biosphere." In 2006 frog watchers positively linked many of the lethal fungal epidemics in the tropics to climate warming.

Nobody knows how the fungus, which lives in water and soil and normally dispatches insects, actually kills frogs. Many scientists now believe that chytrid hampers either oxygen intake or the flow of salts through the skin. There is some evidence the fungus might also produce a deadly toxin. In any case, death usually follows within 12 hours and ponds go silent.

The fungus appears to be a companion of the South African clawed frog. In the 1930s, that amphibian got swept up in the global fertility-test trade. Labs in Europe, North America, and Australia imported the African frog and its freeloading fungus for use in human pregnancy tests. After a sample of a woman's urine was injected under the frog's skin, the frog would ovulate if the telltale hormones were present. The pregnancy trade probably let a few frogs escape here and there (we are not a fastidious species), and soon the fungus was acting like another opportunistic invader dispatching unwary amphibians.

At some point the restaurant trade in bullfrogs, another species that can carry the fungus without harm, helped spread it farther. The bullfrog trade is no idle business. The United States alone imports 1 million bullfrogs from teeming frog factories in Uruguay, Brazil, and Asia every year. Bored restaurateurs and herpetologists have also released the bullfrog in places it doesn't belong. As a result the omnivorous bullfrog has gobbled up the locals and become a fearful invader in its own right. But as the silent carrier of chytrid, it has delivered a double blow to amphibians. Of nearly 6,000 known species of frogs, toads, and salamanders, almost a third are now at risk of extinction. Another 43 percent appear to be in a state of decline. Simon Stuart, who led a recent scientific assessment of amphibian health, summed up all the death and dying this way in 2004: "Since most amphibians depend on fresh water and feel the effects of pollution before many other forms of life, including humans, their rapid decline tells us that one of Earth's most critical life support systems is breaking down."

The breakdown is more advanced in some bodies of water than in others. The future, in fact, can be found in Lake Victoria in sunny Africa. If we don't restrain our ocean meddling by limiting ship traffic, restricting fish factories, or simply cleaning up human and livestock waste, the world's oceans and rivers will all eventually look like the world's largest tropical lake. It's a shallow inland sea about the size of Ireland that once boasted so much ecological diversity that scientists called it Darwin's Dream Pond. But in the last 30 years, a succession of invaders, from the Nile perch to AIDS, have turned the lake into Darwin's worst nightmare, and locals have witnessed the greatest species die-off anywhere on the planet.

The mass extinction took a lot of global prodding. The lake that natives still affectionately call Nam (which means "endless sea") formed about 12,000 years ago and was soon populated by chiclids, colorful little fish the locals call furu. Over time the furu developed into nearly 600 spectacular species that occupied every nook and cranny of Nam's shallow waters.

The furu put on the same sort of dazzling evolutionary fashion show that Galapagos finches displayed for Charles Darwin. Some furu fed off waste while others ate algae. Some crushed snails while others munched plants. Some dined on prawns while others cleaned fish scales. On the shores of Lake Victoria, thousands of native fishermen netted the furu while thousands of women dried, smoked, salted, and grilled them. Even in the 1950s, Lake Victoria still had a sustainable furu economy.

But that all fell apart under the incredulous eyes of Tijs Goldschmidt, a Danish taxonomist. The Mzungu (like all Europeans, he was called a "wanderer") arrived in 1981 to help classify the extraordinary furu. But he ended up witnessing a mass extinction event instead. At first Goldschmidt had no idea why his nets were coming up empty of furu. "Nobody could have predicted in detail the changes taking place in Lake Victoria." But when he started to see huge piles of his favorite fish bursting in the stomachs of obscenely sized Nile perch, he got suspicious.

The Nile perch, a bland-tasting fish native to many northern African lakes and rivers, can grow 2 meters (6 feet) long. The females lay as many as 6 million eggs each in a year. Biologists rank the fish among the world's 100 worst invaders, and it can now be found in Cuba, Asia, and Central America.

Breeding perch in the lake was a British idea. Dissatisfied with the scale and diversity of fishing on Lake Victoria, colonials voted for productivity instead. In the 1950s, they threw in the first Nile perch. As one old woman told Goldschmidt: "The white people put the baby monsters in the lake to help us." It took about 20 or 30 years for the invader to explode. When it did, the Nile perch started to eat its way through one furu species after another. "Every time I thought about it, I was amazed that for the total disruption of the largest tropical lake in the world, nothing more had been needed than a man with a bucket."

As early as 1960, ecologists warned that the invasion could have a disastrous result, but as usual no one listened to ecologists or worried about consequences. Government officials reasoned that bigger alien fish could feed more people than small native ones. So the Nile perch merrily gobbled up about 200 species of furu and took over the lake. When it couldn't find furu anymore, it started eating prawns, and when the prawns got skimpy, it dined on younger perch. When Tanzanians discovered that the invader ate its young, many superstitiously shunned the fish, fearing that they too might become cannibals.

By the 1980s, the monstrous perch accounted for 80 percent of the lake's biomass. "Every ecologist would panic if he saw vast herds of lions in the Serengeti running after the last existing antelope," noted Goldschmidt, but no one panicked on Lake Victoria.

Instead, a new global perch industry merely supplanted the furu economy. In Russian planes, crates of iced perch now fly out of Africa to be served on Japanese and European tables, where eating cannibalistic livestock is old hat. Although the invader has increased nutrient production fivefold, none of the nutrition stays home and half of the children around the lake go hungry. Thousands of women have lost their jobs processing furu for local markets, and most of the perch goes to fish factories elsewhere, run by wanderers from India and Europe.

With no furu to peddle anymore, local women started to sell sex for the scraps of the global Nile perch trade in the 1990s. They'd wait for the fishermen to arrive with big loads of perch and then turn tricks for fish heads or tails. This partly explains why fishing villages in Kenya now have the highest rates of AIDS infection (70 percent) in Africa. "They do sex because they need the money," reports Martin Seemungal in Homa Bay, a dusty town with 35,000 AIDS orphans. In Uganda an entire fishing industry designed to exploit the unexpected fecundity of an invading species may soon be annihilated by the predatory work of an invasive virus. But that's how invaders work: hand in hand.

While the perch rearranged the sex trade and local economy, it also reshuffled the biology of the lake. With the algae-eating furu gone, blue-green algae exploded, and soon the lake experienced regular fish-killing blooms richly fed by nutrient runoff from logging and farming. The water grew so turbid and starved of oxygen that the degradation invited another monstrous invader.

Water hyacinth, a native of South America, is one of the world's fastest-growing weeds. It arrived in Africa as a colonial pool ornament. But in the 1990s, it began suffocating the surface of the lake with mobile mats that choked nearly 90 percent of the shoreline. The Hispanic invader gave the African lake a true Latin touch but made fishing or boating almost impossible. It took an expensive campaign, including plant killers, mechanized hyacinth removers, and the introduction of hyacinth-eating insects, to beat back the invader.

The Nile perch, *Lates niloticus,* has left the region and its 30 million people with a fishing monoculture that serves global markets while ignoring local bellies; it has removed diversity and therefore much resiliency, and it has invited a string of unpredictable catches such as AIDS. In many ways the lake's incredible meltdown reads like some perverse Aesop fable.

Maybe this marine calamity resembles the parable about the too-clever donkey whose scheming leads to disaster. Or maybe the story of Lake Victoria is an avant-garde retelling of the tale about the older man, now a wandering global businessman, with two demanding lovers. As

Aesop told it, the young lover hated white hairs while the older concubine disliked the dark ones. Together the female pluckers turned the man into a bald and unhappy polygamist. The moral? Aesop believed that "ill-assorted companions never come to any good."

The kaleidoscope of diseases now invisibly plying our warming waters are killing diversity with ill-assorted invaders. The ocean, the world's primordial womb, now looks and smells like that man-made soup known as Lake Victoria. That means, of course, that the tide of ocean diseases, like the reach of cholera, will rise day by day.

EIGHT

Climate Riders

"The sheer pressure of our numbers, the abundance of our inventions, the blind forces of our desires and needs, appear unstoppable and are generating a heat—the hot breath of our civilization—whose effects we comprehend only hazily."
—Ian McEwan

Virginia Sherr encountered the dark side of climate change at the age of 55 in her Pennsylvania garden 15 years ago. That's when the healthy psychiatrist got an ugly rash the shape of a bull's-eye on her left leg. Sherr thought nothing of the small black dot that appeared in the center of the red lesion. It just wasn't big enough to be the bite of a tick or spider, she reasoned. Shortly afterward a string of calamitous ailments hit Sherr.

First came chronic fatigue and the abyss of sleeplessness. Then severe arthritic pains invaded her entire body. Restless legs, an irregular heartbeat, and bloated bowels followed. The shrink was stumped. "I had extreme body weakness and severe multiple, specific muscle pains and tenderness that I can only describe as unearthly."

Sherr went from one physician to another looking for answers. Most told her that she looked fine and was just getting older. Although she tested negative for Lyme disease, North America's most underreported and most common infectious disease, the physicians forgot to explain that ordinary blood tests for the disease are notoriously insensitive and unreliable. Years passed and Sherr's bizarre symptoms worsened. The muscles in her hands withered and her body became hypersensitive to touch. Panic attacks, depression, and anxiety became her constant companions. When she no longer had the strength to open the door of her refrigerator in 1997, she realized that her life "was teetering on the brink." Irritability reigned over her emotional life like an unending thunderstorm.

Finally a new family physician identified the invader as a cousin of syphilis: a spirochete with a rambunctious name, *Borrelia burgdorferi*. Sherr then consulted a neurologist, who ordered a stack of highly specific blood and urine tests. To her horror, they showed that her body had been hijacked by "chronic multi-system-disseminated Lyme disease." She also had two other tick-borne ailments: babesiosis, a malarialike disease caused by a parasite that infects red cells, and ehrlichiosis, resulting from a bacterium that wears the body down with flulike aches and pains.

It took two and a half years of antibiotic therapy, but Sherr eventually cleared her blood of the tick-borne freeloaders and is now well. "I understand tick-borne pathogens to be remarkably psychic organisms" that can flourish in any kind of emotional soup, she wrote in 2000. "They wax and wane, submerge and then ambush as if they have minds of their own. I am not surprised that tick-borne diseases have been called fiendish."

Thanks to warming winters and a host of subtle yet intriguing land use changes, tick-borne diseases are making devilish inroads in Europe, North America, and Asia. Since the 1970s, black-legged ticks infected with Lyme disease have expanded from the town of Old Lyme in Connecticut to 47 states, several Canadian provinces, and 30 countries on six continents. In Sweden, Germany, and England, the rising incidence of Lyme disease (also called Lyme borreliosis) and other tick-borne infections is stunning researchers. The U.S. Centers for Disease Control, which admits it does a poor job of tracking the disease, reported that the number of Lyme cases jumped 40 percent from 17,029 in 2001 to 23,763 in 2002. Given that 9 out of 10 cases are not detected, many U.S. experts now suggest that millions of North Americans may be infected.

Because tick-borne bacteria can mimic multiple sclerosis, Bell's palsy, and amyotrophic lateral sclerosis as well as every psychiatric disorder known to the medical profession, many doctors now boldly suggest that 50 percent of chronic diseases may actually be due to undiagnosed Lyme infections. Twenty-five percent of the victims are under the age of 15 and experience devastating neurological symptoms that affect learning. Lyme has become one of world's most common insect-borne diseases in just 30 years. It is to northern peoples what malaria, a mosquito-borne killer, is to southern peoples: a vector-borne menace that gains greater market penetration every day. It is an invisible epidemic and a deadly climate rider with serious ambitions.

But Lyme is not the only germ taking advantage of climate change these days. Warmer weather has opened a veritable Pandora's box of infectious diseases. Some of these climate-sensitive celebrities (and every disease responds to weather as fastidiously as a mountain climber) are novel invaders, such as West Nile virus; others are reliable old killers, such as malaria and dengue fever. The World Health Organization now estimates that at least 7 percent of all new malaria cases can be blamed on warmer weather. But what really worries scientists is the uncontrolled nature of this experiment with our environment. Paul Epstein, an expert in tropical diseases and associate director of Harvard Medical School's Center for Health and Global Environment, has been beating this controversial drum for some time. He believes that warmer weather is already unleashing a wave of nasty infectious surprises. "We've clearly underestimated the rate at which climate would change, and we have underestimated the response of ecological systems to that warming."

Despite the best efforts of oil lobbyists to cloud the issue, climate change is a stunning reminder that bad things happen. Our happy addiction to fossil fuels has produced megavolumes of carbon dioxide and methane gases. These flatulence-like emissions have changed the chemistry of the atmosphere and warmed the planet's surface and oceans by an average of 1 degree Celsius (1.8 degrees Fahrenheit) in the last century. Scientists often refer to the impacts of this subtle shift as "the one-degree factor." But the average is misleading. At higher altitudes and in northern countries, global warming translates into dramatic five- or six-degree-Celsius increases as well as much warmer nights and shorter winters. It also means less snow, more rain, and lots of wacky weather.

Extreme events, such as hurricanes, droughts, floods, and tornadoes, are rapidly losing their status as rare events. (Just ask any insurance company.) Extreme weather is now a global topic of interest and the subject of round-the-clock TV programming. And the show is about to get livelier. Within the next 50 years, scientists predict, temperatures will double or triple as carbon dioxide levels surpass records set 420,000 years ago. Scientists who have studied Antarctic ice cores believe the earth was a very hot and uncivilized place then.

Although business and political leaders have been slow to respond to rising temperatures, the dynamic natural world detests sloth and puts survival first. So its members have begun to shift, slide, and glide to get out of the heat or take advantage of it. Plants, germs, insects, and birds are now doing what they've always done in the face of climate shifts: They're altering their habits or pulling up stakes and moving en masse, like Moses and his people, to more hospitable climes. Flowers are blooming earlier and birds, such as tree swallows, are breeding eight days earlier. Salmon, hornets, barn owls, and robins are turning up in the Arctic, where startled Inuit simply don't have names for them. Ocean creatures along the coast of California are moving north, as are butter-flies, birds, and bats. Off the coast of Norway, fishermen are landing John Dory, a Mediterranean species. Some species, such as the Edith's checkerspot butterflies, have extended their northern range by 60 kilo-meters (100 miles). "While the physics of global warming can be quite subtle," explains Stuart Pimm, a conservation biologist, "it can have major biological effects."

Not all creatures, of course, can adapt or escape the heat in time. Trees aren't fast runners; they tend to dry up and become kindling for massive fires or beetle invasions. Amphibians, now being decimated by fungal invaders, are also having trouble finding their sneakers. Tiger beetles and dragonflies may not be able to pack their bags fast enough either. Some unfortunates in the Arctic and in highland regions simply have nowhere to go. The colorful quetzal bird of Guatemala, for example, lives in the cloud forests of Central America. But as they dry up, the quetzal will soon find itself homeless, replaced by toucans. Several studies now warn that climate change could burn up anywhere between 20 and 50 percent of the earth's 10 million plant and animal species. "We have rather little idea as to quite what this is going to do to the functions of biological ecosystems," notes a British conservation biologist, Chris Thomas. So climate change has set the world's creatures buzzing on various evolutionary and unpredictable paths with all the careless abandon of a blind child whacking a hornet's nest with a stick.

Bacteria, viruses, fungi, and protozoa, the globe's oldest and most seasoned climate adapters, are also reading the logbook of global warming. And for many, it's mostly good news. In many regions, climate change has banished cold winters, one of the planet's most effective disease filters and pest killers. Thanks to milder and shorter winters, a host of disease-carrying insects, including ticks, snails, and mosquitoes, are living longer, eating more, and reproducing faster. In fact, the one-degree factor makes it possible for many species to double their numbers in half the time, a novel evolutionary trick humans have yet to master. Two of the world's most serious mosquito-borne infections, malaria and dengue fever, not only quicken their pace in warmer weather but also expand their hunting grounds. The Australian government, for example, predicts that the number of people at risk from dengue infection will grow from 300,000 to 1.6 million as things heat up within the next 50 years. So climate change can take a tropical menace and turn it into a temperate one.

Moreover, climate change tends to dry out landscapes or flood them; either extreme creates ideal conditions for viral and bacterial flare-ups. Hantavirus, a nasty lung collapser transported by mice, requires heavy flooding in order to get into shape to score human fatalities. Cholera, a waterborne bacterium that can dry up a human in 12 hours with relentless diarrhea, knows how to take advantage of extreme events, such as flooding and warm waters, and is still enjoying its longest and most wide-ranging pandemic in recorded history. Neisseria meningitis, a common brain sweller in Africa, depends on drought to spread on billowy clouds of dry dust. In 1994 it killed 100,000 people, an all-time record. Certain diseases, such as typhus and avian cholera, both cold-weather lovers, appear to be on the retreat. But for most diseases, milder weather encourages population bonanzas.

Last but not least, warmer weather and extreme storms unsettle human surroundings and infrastructure. Hurricanes, floods, and drought not only batter people and their institutions but stress the land as well. Many places already degraded by exuberant tree cutting, desperate dam building, and garish city making no longer have the resources to deal with hotter and wilder weather. In fact, disturbed land behaves much like a human with a busted immune system: It simply becomes more vulnerable to invasions. After record rainfalls submerged the crowded streets of Mumbai in 2005, killing 1,000 people, cholera, typhoid, and leptospirosis (caused by a spirochete carried in rat urine) all came calling.

When Paul Epstein and a number of other muckracking scientists first raised the specter of a warmer world being a sicker one a decade ago, a lot of colleagues pooh-poohed the notion. They called Epstein's hypothesis reductionist and argued that it was impossible to separate the bad work of something as complicated as climate change from human-driven landscape makeovers and the ever-escalating traffic of goods, people, and living things. Many also claimed that humans would surely adapt to higher temperatures with smarter health campaigns, more vaccinations, and better pest control. In other words, if the mosquito populations started to explode outside, then people would just stay indoors and watch more TV. Others ranted that the problem wasn't climate change at all. They pointed fingers at the collapse of public health systems as well as startling epidemics of drug-resistant parasites.

Although some of these arguments remain valid, recent studies strongly suggest that climate change has added more fuel to an already raging fire with all the stealth of a pyromaniac. In 2002 several American ecologists sifted through hundreds of scientific papers for the prestigious journal *Science* and found an unnerving number of climate-driven epidemics affecting plants, trees, wildlife, and man. Every type of pathogen from viruses to fungi appeared to be involved, and every type of species seemed to be under attack. Birds, oysters, orange trees, lions, ferrets, deer, lobsters, and wheat were all experiencing sudden die-offs or debilitating illnesses because of disease. The animal ecologist Richard Ostfeld found the sheer number of outbreaks simply shocking: "We don't want to be alarmist but we are alarmed."

The study painted a Hieronymus Bosch picture of a warmer and more humid earth sopping with bacterial and viral growth. In Hawaii warmer temperatures have invited avian malaria to move up mountains, where it is dispatching the honey creeper, a resplendently colored song-bird. Fungal disease has taken advantage of balmy sea temperatures to chew up sea fans in the Caribbean. Across California an unrepentant fungus has been felling oak trees in less than four months flat. An oyster-killing protozoan parasite nicknamed Dermo has expanded its range along the U.S. eastern seaboard thanks to warmer waters. In the last 14 years, the Dutch elm disease fungus has eaten more trees in England in the warmer years than in the colder ones.

In the last 10 years, the warmest on record, just about every insect-borne disease from malaria to bluetongue fever has zealously expanded its territory. So many climate riders are galloping forth that the comprehensive *Science* study missed a whole bunch. Hurricanes have helped spread the citrus canker throughout Florida's fruit monocultures, with billion-dollar losses. On Vancouver Island a tropical fungus native to Australia (*Cryptococcus gattii*) has become 37 times more infectious in a temperate zone, thanks to drier than normal summers. The deadly invader, which has sickened more than 100 people and killed four, is also spreading. England has become such a balmy island that the stink bug, a voracious crop eater, has quietly established several colonies in London. In 1959 bug experts firmly predicted that cold would keep the stowaway an occasional visitor. But climate change made liars of them.

To many ecologists these separate tales add up to trouble. "What we found were striking patterns of climate warming and spread of disease and greater incidence of disease," concluded Ostfeld in the *Science* study. Andrew Dobson, an evolutionary biologist at Princeton University, added that climate-driven plagues were "a much more scary threat than bioterrorism." To Epstein, these and other findings merely confirmed his darkest fears: "Volatility of infectious diseases may be one of the earliest biological expressions of climate instability."

Of all the climate riders making headlines, Lyme disease probably illustrates that volatility the best. Not only is it one of the world's most ignored invaders, but it has arguably done a lot more damage to human health than West Nile and SARS combined. More important, Lyme shows how quickly a stealth invader can insinuate itself into the fabric of modern life by taking advantage not only of warmer weather but also of suburban monocultures, simplified forests, and an abundance of deer, mice, and medical arrogance. James Burranscano, one of the continent's few Lyme experts, recently summed up the seriousness of the matter before the U.S. Senate: "The truth is that Lyme is the fastest growing infectious illness in this country after AIDS, with a cost to society measured in the billions of dollars."

The Lyme invasion ostensibly began in 1976 when moms in the town of Old Lyme, Connecticut, raised hell after local doctors wrongly diagnosed their teenage sons and daughters with debilitating bouts of so-called rheumatoid arthritis. Some of the afflicted couldn't even walk. But the parents knew that fever and aching joints don't suddenly appear in clusters of children. Nor did they accept arthritis as a normal condition for young folks. And they were quickly proven right. In 1981 Willy Burgdorfer, a Swiss-trained tick expert at Montana's famed Rocky Mountain Laboratory, ground up some deer ticks from the eastern United States and identified a bacterial spirochete called *Borrelia burgdorferi*.

Borrelia burgdorferi comes from a vast bacterial family that has been causing trouble for centuries. *Borrelia's* immediate cousin is the microbe that causes syphilis, a venereal disease that got the better of the medical profession and many Casanovas. (The two bacteria share many of the same proteins.) Syphilis earned the moniker of "the Great Imitator" or "the Great Pretender" for its chameleonlike qualities. What starts as genital sores evolves into a generalized rash followed in later stages by heart disease, blindness, paralysis, and insanity. Psychiatrists have built entire careers on the puzzling behavior of their syphilitic patients. The brilliant physician William Osler once remarked, "He who knows syphilis knows medicine." If Osler were alive today, he'd probably say the same about Lyme disease.

While syphilis begins with promiscuity, Lyme begins with the humble tick. Wise old Aristotle once described these creepers as "disgusting parasitic animals," and no one has really improved on that description. The eight-legged bloodsuckers belong to the same family as mites and spiders and come in more than 850 species, some no bigger than a poppy seed. (Mosquitoes, by contrast, come in 3,500 types.)

Without a good draft of bird, mammal, or reptile blood, the tick can't grow, mate, or procreate.

As blood drinkers, ticks have developed a demanding lifestyle. Immediately after birth, the larvae need a blood feast in order to molt and turn into nymphs. Nymphal ticks, the most common human biters, require another sanguine drink before molting into adults. Some species wait in grass or leaf litter up to four years to find an unwitting blood donor in the form of a mouse or a shrew. Blind as bats, ticks use carbon sensors in their rear legs to detect sources of blood. Once engorged, ticks fall off their hosts and mosey about on moist soil and leaf litter. A female can lay 3,000 to 6,000 eggs at one go, illustrating once again that the world's evolutionary fate lies in the hands of the irrepressibly fertile.

Like mosquitoes, ticks are bona fide members of the climate change club. The one-degree factor can mean the difference between millions and billions of ticks. When temperatures shot up 1.6 degrees Celsius above average at Mount Tamborine in Australia, daily tick infestations on cattle increased from 200 to 1,400 per animal in a year. "This illustrates the important point that there is the potential for disproportionately large increases in the rates of disease transmission with small increases in temperature in cool-limited habitats," wrote the Australian entomologist Robert Sutherst in a lengthy 2004 paper on climate change and human vulnerability to insect-borne diseases. Milder winters guarantee the survival of more tick eggs, and longer summers enable ticks to expand their range. More moisture in the form of rain instead of snow is also a tick-positive development. Cushy winters also make it easier for deers and mice, both mammals swimming with spirochetes, to reproduce in greater numbers.

Ticks aren't particularly picky about who or what they dine on. In the early stages of their life cycle, some species, such as *Ixodes scapularia*, favor mice, squirrels, and chipmunks in eastern United States. Nymphs and adult ticks tend to dine on larger hosts, such as deer, hikers, gardeners, forest workers, and children. On the west coast *Ixodes pacificus*, another Lyme carrier, sucks the blood of lizards, snakes, coyotes, and robust Californians. Migrating birds, from puffins to seagulls, also make good blood meals and have probably played a crucial role in expanding tick colonies. While making withdrawals, larval ticks invariably pick up a variety of bacteria, viruses, and protozoans, which the nymphs then pass along to the next blood donor, whether it be mammal, bird, or man. In the last two decades, every tick-infested country has reported exploding tick densities and, with them, more tick-borne diseases.

Humans aren't the only ones feeling the brunt of the global tick invasion. In Scotland ticks are becoming so numerous between May and August that they are killing grouse in record numbers with a deadly virus. In Canada infestations of more than 50,000 ticks per moose have driven the stately animals crazy enough to rub off their hair and have resulted in massive die-offs during warm winters. Throughout North America pet owners are finding their dogs, llamas, horses, and cats crawling with ticks and associated ailments. (Dogs can get Lyme disease.) In Kansas, Michael Dryden, a professor of veterinary parasitology, admits that tick experts can no longer keep up with the explosion. "Tick research is lacking today because historically it was not as serious an issue as it is today." (Invading ticks may also add to the mayhem: A new immigrant introduced through Japan's black market trade in beetles now threatens to kill off that nation's indigenous beetles with a bacterium that causes their limbs to rot and fall off. Finding Asian and African ticks on imported lizards, toads, and other reptiles is now routine in North America.)

Entire books have been written about tick-borne fevers, and for good reason. Ticks can deliver as many as 60 different pathogens to livestock and more than 60 viruses and 11 bacterial diseases to people. In Australia cattlemen worry that disease-laden ticks will soon double their range, thanks to milder winters, which will double economic losses in the cattle business. Africans, a people already scarred by tick-borne diseases, continue to worry over East Coast fever and heartwater fever. An alarming human epidemic of tick-borne encephalitis, a virus that causes tremors, blinding headaches, and mood swings, has Europe in an uproar. In North America ticks now carry as many as eight identified human invaders, including Rocky Mountain spotted fever, tularemia, and of course the infamous Lyme.

Borrelia burgdorferi is an invader prodigy: It is as stealthy as Rasputin and as ruthless as Genghis Khan. After entering the human bloodstream, Bb quickly embeds itself in muscles, the heart, and brain tissue or in any part of the body that antibiotics have difficulty reaching. The bacterium then changes its shape from a corkscrew spirochete to a cyst, making it adept at avoiding the body's immune system as well as most blood tests. (The highly insensitive serologic blood test for Lyme misses 40 percent of the cases.) So, like syphilis, Lyme can hide, reappear, and mess with the brain in a spectacular number of ways over a long period of time. To make matters worse, Bb boasts an uncomfortable diversity: five subspecies and nearly 300 strains worldwide. Last but not least, *Borrelia,* like many competent invaders, rarely travels alone. A tick bite in North American delivers not only Bb but often also a good dose of the protozoan *Babesia* and the bacterium *Ehrlichia,* another exploiter of the immunocompromised.

The invader also doesn't seem content to be mere tick cargo. Live Bb spirochetes have been recovered from gnats, fleas, mosquitoes, urine, tears, semen, and blood. Infected mothers have passed on the spirochete to their children. Some doctors, alarmed by the number of cases in areas with limited tick activity, suspect that humans can even pass on the spirochete through blood transfusions and sex. As a consequence some doctors treat the apparently healthy spouses of infected patients along with their symptomatic partners.

Like most chronic ailments, Lyme disease is a diagnostic nightmare that has confounded the medical establishment. Some patients produce a red ring rash, but 60 percent don't. Many complain of fever, muscle aches, stiff neck, and chronic fatigue within a week of being bitten, while others go symptomless. (More than half of all Lyme victims don't recall being bitten by ticks, and one in three people dined on by nymphs never knows the vampire was there.)

Within months heart problems ranging from chest pain to irregular heartbeat may develop, as well as arthritis and deadly fatigue. Two to eight years after infection, many patients, as Virginia Sherr's story documented well, become overwhelmed with a rainbow of neurological symptoms including brain fog, paranoia, facial nerve palsy, meningitis, and garbled speech. In children—a quarter of all infections dog the young—the spirochete can dramatically affect learning by inducing fatigue and inattentiveness. Some children miss up to 100 days of school when chronically infected.

It is common for physicians to misdiagnose Lyme disease as flu, stress, or a bad case of modern hysteria. Patients are routinely told that they have multiple sclerosis, chronic fatigue syndrome, or a host of other vague syndromes that serve as a catchall for medical ignorance. One 19-year-old German patient, with classic neuroborreliosis, behaved like a catatonic and paranoid madman with delusional ideas of persecution. He was headed for psychiatric care until doctors isolated *B. burgdorferi* and prescribed antibiotics. Unfortunately many patients can visit dozens of doctors before being properly diagnosed or ending up in asylums.

The association between *Borrelia burgdorferi* and chronic disease is alarming and poorly studied. Patients with chronic fatigue syndrome, multiple sclerosis, and Alzheimer's have all been found teeming with Lyme spirochetes. MS and schizophrenia share the same geographic distribution as Lyme-carrying ticks. MS patients treated with antibiotics capable of destroying Lyme bacteria have experienced fewer neurological attacks or nearly complete recoveries. One Swiss researcher, Markus Fritzsche, has noted that the birth dates of individuals who develop schizophrenia mirror the distribution of ticks nine months earlier at the time of conception. His radical hypothesis: that prenatal infection by Bb may lead to a preventable brain disorder.

Whenever a new and chronic disease invades the medical consciousness, controversy follows. One group of doctors maintain that a simple three- to four-week course of antibiotics will do the trick for Lyme disease and that relapses are impossible. Yet not one medical study actually supports this rushed therapy. A small cadre of Lyme specialists now argue that without a high-dose course of antibiotics for 6 to 12 months, the spirochete will survive to reappear in even more debilitating guises. Thousands of patients concur. In fact, the existence of hundreds of Lyme support groups, rowdy legislative inquiries, and tens of thousands of dissatisfied ill patients strongly suggest the medical profession has once again been sorely outwitted by a spirochete.

The origin of Lyme is also the subject of more debate and three distinct theories. European doctors clearly recorded its arthritic ravages on that continent as early as the 1880s. So the agent is not new. Willy Burgdorfer, the discover of the spirochete, now believes that European deer imported in the 19th century may have brought over both the deer tick and the bacterium. But no explosion occurred until the deer, the ticks, and the spirochetes found the opportunities for riotous biological mergers toward the end of the 20th century.

Others theorize that a constellation of ecological changes gave the tick and the spirochete a big break. American pioneers did a solid job of cutting down old-growth eastern forests and exterminating deer and other creatures 400 years ago. A rebound didn't really happen until late in the 20th century when trees, deer, and mice returned in heavy numbers. At the same time, suburban developers planted more and more homes in the shade of leaf and limb. "We made our own sickbed and then we had to lie in it," explained one doctor.

The third possibility is stranger than fiction. Just off the coast of Connecticut, the epicenter of the Lyme epidemic, lies Plum Island. This U.S. biological warfare facility, which has been run by the Pentagon, the Department of Agriculture, and now Homeland Security, has been playing with bacteria and viruses since the 1950s. In its early years, it earned a reputation for being frighteningly insecure, unsafe, and leaky. During the 1950s, it also employed Erich Traub, a former Nazi scientist. One of Traub's specialties was infecting beetles with viral and bacterial agents and dropping them out of planes on unsuspecting Russians. The island, which covers 340 hectares (840 acres), teems with deer and migratory birds. Wild fowl, of course, are fantastic transcontinental tick couriers.

U.S. scientists have admitted a few shameful facts. Researchers at the island nurtured a tick colony in the 1960s and eagerly investigated the possibilities of transmitting heartwater, bluetongue, and African swine fevers to animals via infected ticks. In 1969 Donald MacArthur, a Pentagon biowarrior, told Congress that his peers on Plum Island were actively working on "incapacitating agents" and testing them on animals. An incapacitating agent "imposes a greater logistic burden on the enemy when he has to look after the disabled people," explained MacArthur. The first teenager to come down with Lyme disease lived next to the ferry dock that delivered scientists to the island laboratory. To Michael Carroll, the author of *Lab 257,* a damning investigation of doings at Plum Island, these coincidences strongly suggest that Lyme just might be an accidental man-made epidemic: "Wild birds and deer contracted Bb from ticks impregnated with a myriad of exotic germs studied on Plum Island in helter skelter, haphazard conditions."

Whatever its birthplace, Bb aggressively colonized tick populations in fragmented forests abounding in deer and mice. And it did so with all the determination of the zebra mussel. In some parts of New England, it now infects as many as 400 out of 100,000 people, the highest infection rate in the world. In New Jersey it is not uncommon to find entire families that have been infected or disabled by Lyme disease. The bacterium is now present in ticks in 47 states and is entering Canada via songbirds. It has already colonized Ontario, Manitoba, Nova Scotia, Quebec, and New Brunswick. A 2005 study suggests that climate change will boost the amount of suitable habitat for ticks in Canada by

213 percent by 2080. In fact, wherever you find fragmented forests, deer, mice, ticks, and people gardening in their backyards, you can find Lyme disease.

Suburban monocultures have clearly given Lyme a major boost. The old-growth forests of New England, for example, used to harbor a wealth of species including foxes, opossums, lynx, and various fowl. Most of these creatures aren't terribly competent carriers of Lyme. But the region's suburban neighborhoods don't reflect the diversity of older habitats and favor instead armies of white-footed mice and eastern chipmunks. Ninety percent of the ticks feeding on these creatures pick up Lyme. If the feeding options for the tick were more diverse, many scientists suspect a dilution effect would greatly reduce the number of ticks carrying the disease, thereby lowering the human Lyme infection rates. But most of the U.S. northeast as well as much of suburban Europe offers ticks a highly concentrated menu of infected rodents.

So a land that seems hospitable to humans can become a paradise for ticks and, with warmer weather, a veritable Valhalla for arthropods. California now calls Lyme disease a "public health epidemic." In Minnesota, where the Department of Health actually runs a decent surveillance program, the number of cases reached 867 in 2002, a whopping 86 percent increase over the previous year. Meanwhile one of Lyme's dangerous fellow travelers, ehrlichiosis, has increased by 60 percent in the state. University of California researchers recently took a stroll in the woods and found that hikers had a 30 percent chance of picking up a tick every time they rested on a log, picked up wood, or sat against a tree. A 2003 report on animal-borne epidemics by the Trust for America's Health concluded, "Lyme disease has become a permanent part of America's public health landscape. It provides a warning and example of how an apparent state or regionally-centered problem can grow to become a national problem."

The tick invasion is not confined to the United States. In Europe, Lyme infection rates now rival those of North America. In the last decade, the percentage of nymphs carrying different strains of *Borrelia* has jumped (from 13 percent to 21 percent) in a nature reserve outside of Bonn. Thanks to such explosions, more than 40,000 Germans come down with Lyme disease every year. Researchers blame the rising tick densities and spirochete prevalence on warmer weather and milder winters, noting that climate change benefits ticks, deer, wild boar, and migrating birds alike. Another study found that 5 percent of Germans bitten by ticks picked up nasty babesiosis infections. England, which claimed to be Lyme free in the 1980s, now reports more than 300 cases a year. Austria and Croatia are positively crawling with ticks and report the highest rates of infection in Europe. Lyme disease increased 15 percent in Hungary between 1999 and 2005, and 20 percent of ticks in the Omsk region of western Siberia now transport the spirochete. Scientists in Lorraine, France, blame a dramatic increase on the way humans have carelessly modified the landscape. Suburbs, it seems, just invite a troublesome triumvirate of ticks, deer, and mice.

Ticks, however, don't always need to feed on big mammals to pick up bacteria and are definitely going urban. In Finland tick watchers found densities of 36 critters per 100 square meters in Helsinki and concluded, "Lyme borreliosis can be contracted even in urban environments not populated with large mammals such as deer or elk." In other words, the ingredients for an urban epidemic were simple: ticks feeding on infected mice and shrews. In an urban park in Prague, as many as 10 percent of the ticks now carry the spirochete. In suburban Switzerland, 12 out of 13 hedgehogs carry ticks infected with Lyme. A 2003 report on vector-borne disease in Europe noted the obvious about all this Lyme disease activity with a medical solemnity: "On the whole it must be concluded that the increase is real."

Nor is Lyme disease the only tick-borne invader crowding doctors' offices or climbing up the health charts. Tick-borne encephalitis (TBE), a virus that can swell the brain and leave older victims with brutal neurological damage, including deafness, is infecting more Europeans than ever before and colonizing new ground. Goat's-milk drinkers are also at risk. Ticks shed the virus in goats, which secrete it in their milk, a fact that has led to prohibitions on drinking raw goat's milk in many countries.

In the Czech Republic, where infection rates have jumped from 300 to 700 new cases a year, the tick vector has even taken to climbing to higher altitudes, another trick encouraged by climate change. In the warmest decade on record, Estonia, Latvia, and Russia have all reported stunning increases in TBE: as high as 2,200 percent in a single year. Sweden's caseload has tripled since the 1980s. In Lithuania the number of infections doubled from 200 to 418 in the last decade. Noted one medical reporter: "Man and tick have been living side by side for ages but the more destructive interaction only began some years ago when ticks brought diseases unknown to us." Now ticks are even biting folks in parks, gardens, and "the green areas of the city center."

The Swedes, Europe's most passionate outdoors lovers, have fingered climate change as a key driver of Europe's TBE explosion. Elisabet Lindgren at Stockholm University studied the incidence of TBE infections over a 36-year period in Stockholm County and found that things were definitely heating up. In 1998 she reported that infection rates had shot up threefold in 1994, after five consecutive mild winters and seven early springs. Two years later she looked at tick density in 20 different districts by asking Swedes to count the number of bloodsuckers on their cats and dogs. She again found the highest density of ticks (as many as 100 per animal) in districts with milder winters and extended spring and fall seasons. If the climate becomes warmer, she warned, "ticks may spread into regions at higher latitudes and altitudes as well as become more abundant in established regions."

Sarah Randolph, a tick expert at Oxford, also predicts that global warming will drive both ticks and TBE further north into Scandinavia. In addition to tick-friendly weather, she attributes the astounding increases in tick-borne ailments to the collapse of Communism. She suspects that the fall of tyrannical regimes invited more people to go outside to enjoy the great outdoors or to forage for berries and mushrooms. (Warmer weather might have provided the same invitation.) Climate change combined with our penchant for fragmented forests in suburban communities and our love of dachas and deer has guaranteed that our children will be struggling with a revival of tick-borne diseases for years to come.

T icks aren't the only pests on the move. Skeptics of climate change and the one-degree factor should consider the progress of another bug: the lowly mountain pine beetle. It's about the size of a rice kernel, and it's unsettling Canada's forest industry the way ticks and Lyme have rattled public health authorities in New England. In fact, the beetle has already become the fiercest competitor ever faced by multinationals that hew wood. Originally found only in British Columbia, the mountain pine beetle, or MPB for short, has taken advantage of milder winters and warmer nights to go farther afield. Thanks to its insatiable appetite for mature lodgepole pine, the MPB is now the notorious star of several government and industry Web sites detailing its mega-predations. The tree slayer may even chew its way across Canada's boreal forest, one of the world's last intact woodlands.

A few shocking numbers explain the beetle's trip from anonymity to continental notoriety. Since 1996 the population of MPBs has doubled every year. Trillions of beetles have chewed through a forest area nearly the size of Poland. In the process the beetles have destroyed 10 million hectares (25 million acres) of lodgepole pine, British Columbia's primary commercial tree species. That's $18 billion worth of wood, or enough lumber to build 5.2 million homes. Damage to the province's economy is estimated at $6 billion and could easily top $50 billion within a decade. Millions of dying trees have also robbed parks and watersheds of vital scenery and erosion guards.

Although in ordinary conditions the MPB has an essential role in pruning aging and crowded forests, the pine beetle owes its current pariah status to climate change. Warmer winters (a 2.2-degree-Celsius increase in the British Columbia interior over the last decade) have encouraged the beetle to multiply by 4,000 percent in some regions. And a misguided policy of fire suppression that has left more pines standing than ever before in the province's history has given the MPB unprecedented dining opportunities: a vast contiguous table of delectable pine. Nor is the MPB the only beetle taking advantage of sissy winters and drought-stressed trees. Warm winters in Alaska unleashed a spruce beetle plague that mowed down nearly 30 million trees on the Kenai Peninsula. "It slowed down," the forest ecologist Edward Berg explained, "only after they had literally eaten themselves out of house and home." Other infestations have felled millions of trees along the Rocky Mountains, creating potent fuel for runaway wildfires. In 2003 a bark beetle attack killed 90 percent of the piñon trees in some northern New Mexico forests. The MPB is poised to join these wrecking crews and has already invaded American forests.

Like most tree-eaters, the MPB shares the hardy characteristics of a fertile invader. As many as 1,000 bugs swarm a tree, burrow under the bark, suck up all the nutrients, and lay their eggs. A year later the needles of the lodgepole turn orangey red and the tree dies. Several million bugs can infest just 1 hectare (2.5 acres) of forest. "The numbers are so staggering it's impossible to describe them," says Doug Linton, a bark beetle specialist at Vancouver's Pacific Forest Centre. The beetles even leave behind a calling card: a blue-staining fungus. The Japanese, major importers of Canadian pine, don't like stained wood.

Scientists fear that if warming trends persist, the beetle will invade higher elevations and eventually spread east through Canada's vast boreal forest, which stretches from Yukon to Labrador. "We can't stop them," adds Linton. Without some minus-40-degree-Celsius (minus-40-degree-Fahrenheit) deep freezes in early winter or later spring, the MPB, which has already gobbled 10 percent of the province's pine, could chew its way through 80 percent of all mature pine trees in British Columbia by 2013. The tree-eating MPB has become just as dramatic a signature of climate change as bloodsucking ticks in Sweden. Nor is it alone. The U.S. Forest Service is discovering as many as five new pathogens or bugs every year introduced by trade in raw wood and wood packing materials. The agency sadly describes forest health as marginal and warns that many tree species are becoming "functionally extinct." The MPB just may be a sentinel warning of worse things to come in our forests.

W est Nile fever, another sunbather of an illness, has broken a different set of records in urban areas. By now most North Americans know that it is something out of Africa and a good reason to buy mosquito repellents, netting, and other defensive paraphernalia during the summer. Since its first dramatic appearance in New York in 1999, the invader's spectacular spread has, as one group of researchers put it, "reflected an unfortunate confluence of viral promiscuity and ecological diversity." Andrew Spielman at Harvard University, one of the world's foremost mosquito monitors, says that the advent of West Nile encephalitis in the United States is "an event that has changed urban life."

West Nile used to live in Uganda, where it passed largely unnoticed as a mild childhood infection. Virologists got interested in it in the 1950s because it killed mice and monkeys but not humans. When American doctors later experimented with the virus in a bold attempt to cure cancer patients, they discovered that the virus could cause swelling of the brain with deadly precision.

Back in Africa, U.S. military researchers found West Nile virus everywhere they looked in rural Egypt. Most adults carried antibodies to it, as did most livestock and birds. As an omnipresent native, the virus seemed to be just a stable part of the neighborhood. But once it started gambling abroad, traveling with migratory birds, it caused some trouble: weeks of fever and "feeling crappy" in Montpellier, France, and dead horses in the Rhone River delta. It also poked its head into several old folks' homes in Israel in 1957, leaving four dead. (Migrating birds routinely stop over in Israel on their way to wintering grounds in Africa.)

In 1996 the virus emerged on the marshy floodplains of Bucharest as a fatal brain sweller. An epidemic overwhelmed the hospitals and filled beds with adults complaining of the ugliest of flu symptoms. Ten percent of the infected died from encephalitis, tremors, and convulsions. Disease detectives eventually traced the outbreak to a strange confluence of events. Drought had failed to flush out the city's sewer system. Birds then arrived with the virus. The larvae of *Culex pipiens,* a major West Nile carrier, wriggled out of stagnant pools as well as pit toilets and wet basements of the grim block housing favored by Communist architects. The mosquitoes fed on birds and then on the people of Bucharest. Sometimes, notes Robert Desowitz, a discerning U.S. epidemiologist, it's not the big bang that gets us but "a package of small subtle perturbations."

North America's West Nile package arrived during a hot and dry New York summer on a traveler from the Middle East (most likely Israel), a wayward mosquito, or an infected bird, perhaps in the pet trade. The strain looked and acted exactly like one that had devastated Israel's goose factories the year before, with 20 percent die-offs during a blistering heat wave. (Just two years later a highly virulent strain infected hundreds of Israelis of all ages and killed 35 people.)

The African invader quietly announced its arrival in the United States by killing thousands of American crows, a favorite mosquito dinner, in June and July. Then winged biters started to draw blood from New Yorkers instead of avians, sending a cluster of fevered seniors to one community hospital. After four patients died with startling neurological symptoms, including paralysis, one physician became alarmed and started making phone calls.

Next a variety of New World bird species, including Chilean flamingos, a Guanay cormorant, and a bald eagle, dropped dead at the

2005 WEST NILE VIRUS ACTIVITY IN THE UNITED STATES
(Reported to the CDC as of February 14, 2006)

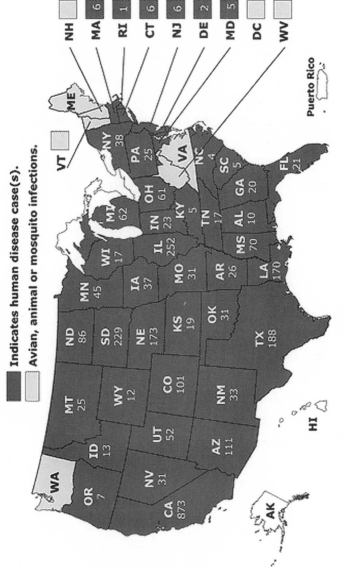

Indicates human disease case(s).

Avian, animal or mosquito infections.

West Nile virus conquered the United States faster than smallpox.

Source: Centers for Disease Control. http://www.cdc.gov.

Bronx Zoo. Tracey NcNamara, a veterinarian and the zoo's head pathologist, started to put two and two together. "The birds were sentinel species. They were waving the red flags for months—but they were just a bunch of crows." McNamara couldn't get the authorities to do the proper tests or to listen to a vet's hunch that the human cases and the bird deaths were related.

Another month passed before the Centers for Disease Control finally conceded that a hostile Old World immigrant had arrived on our shores. If West Nile was a test of our public health surveillance system's ability to deal with even deadlier viruses, it demonstrated a failure "to expect the unexpected." A telling report on the outbreak by the General Accounting Office later found poor communication, inadequate monitoring, and great uncertainty about "what to report, when and to whom." If it had not been for one persistent physician and one angry veterinarian, the invasion might have passed unnoticed for another year.

Since 1999 West Nile has easily become North America's most successful biological invader as well as a sloppy excuse for panic spraying of insecticides. (New York even "scourged" one area in October, long after the end of mosquito season.) In just five years, a modest invasion confined to an 80-kilometer (50-mile) radius of New York City has gone continental and now threatens Hawaii and all of the Americas. What initially emerged as 62 fever cases with seven deaths now counts an average of nearly 5,000 cases with 300 deaths a year (still tiddlywinks compared with the ravages of malaria or dengue fever). The virus now has a gigantic fleet of 30 mosquito species at its disposal and freely swims in 150 bird species. One of the fever's most competent spreaders has been the common house sparrow, an English invader naively introduced to New York City in 1851.

With so many friends, Africa's latest import obviously plans to stay a while. So it's best for young urban folks (like Lyme, West Nile is an urban problem) to get bitten and thereby immunized. For the elderly and immunocompromised, the virus remains another unkind insult delivered by the blind gods of globalization. But over time, as the virus sweeps through human populations with the aid of dry spells, it should become a minor multinational irritant. That hopeful prognosis could change, warns Robert Desowitz, "if it mutates or is made to mutate." Then, of course, "we are in for some big trouble."

No one really knows how climate change aided the West Nile virus. It appears to have taken advantage of warm winters followed by dry hot summers, but the correlation isn't perfect. Canadian researchers have noticed that heat matters. When summer temperatures rise above 30 degrees Celsius (86 degrees Fahrenheit), the infection rate in mosquitoes quickly reaches 90 percent in 12 days, a feat that might take 30 days in colder weather. So climate has put this tropical fever in a driver's seat, but it is only one of several factors propelling the speed and distance of the car.

Dengue, or breakbone fever, already a racing menace because of urban growth and the used-tire trade, may also be given a boost by climate change. Like West Nile, it probably originated in Africa or the Near East. But thanks to the ambitious slave trade and other European enterprises, the virus and the mosquito that carries it, *Aedes aegypti,* a dawn and dusk feeder that goes for the ankles, went global 200 years ago. After it had terrorized the Caribbean and Philadelphia with several big epidemics, people got used to it. Until the 1950s, dengue generally sent its victims to bed for several weeks with an assortment of bone-cracking aches and pains. But in the Philippines, several strains of the virus caught up with one another during World War II and started playing havoc with the immune system, causing a fatal bleed-out known as dengue hemorrhagic fever. The new strain came with a 5 percent death rate. Urbanization and the collapse of mosquito-control programs, particularly in the Americas, gave DHF an unlimited passport.

The used-tire trade, a perfect nursery for mosquitoes, then extended dengue's global reach. In the 1970s, companies in the United States started to import containers full of tires from Asia and Japan for retreading. (The number of steel containers arriving in U.S. seaports has doubled to 33 million since 1990, and even empty ones ferry both live and dead pests.) By the next decade, Houston, then the globe's retreading center, was importing more than a million tires a year. At some point a load of water-filled tires arrived with a big brood of the Asian tiger mosquito. In addition to being a fierce daytime feeder, *Aedes albopictus* is also a fine dengue carrier. When the entomologist Paul Reiter found the highly aggressive East Asian immigrant in a mosquito trap in a Memphis graveyard in 1983, he knew he had found a heap of trouble. The tiger, which lives up to its name, can carry more than 30 different viruses, including such tropical delights as yellow fever.

In the last two decades, the tiger has performed what one researcher calls a "Darwinian blitzkrieg," crowding out *Aedes aegypti* and occupying a lot of territory. It is now a well-established resident of southeast Texas, home to the U.S. used-tire trade, as well as of 25 other states. The tire trade also gave the Asian tiger mosquito a lift to Albania, Italy, France, Nigeria, and Brazil. The tiger has even become Italy's most important mosquito pest. In fact, the tiger has moved into both tropical and temperate cities, where it has found unlimited housing in the forms of crushed tin cans, plastic bags, and the other detritus of city life.

Dengue's recent progress reads a bit like a "one damn thing after another" joke. The used-tire business moved dengue's most competent vectors about the planet for free. At the same time, travelers infected with dengue spread four distinct viral strains, including the bloody DHF, all over the tropics. Urban growth in the tropics has since given both the disease and its carriers ample housing and employment. As a consequence dengue has become almost as fearsome as malaria. Prior to 1970, only nine countries had to worry about DHF. Today it occupies 100 countries and has placed more than 2 billion people at risk. In the 1950s, the WHO recorded about 1,000 cases a day of DHF; today it registers more than 600,000. Throughout Asia, dengue is now the number one hospitalizer of children.

Although dengue has yet to land on North America's shores in serious quantities, climate change may tip the scales. Higher temperatures can accelerate the spread of the disease. There is good evidence that warmer weather has already increased the range of dengue from tropical cities, like Havana, to subtropical ones, like Buenos Aires. Like many mosquitoes, the tiger breeds faster as the temperature gets higher. With dengue outbreaks now routine in the Caribbean (the $12-billion cruise industry is not pleased) and as many as 3,000 travelers returning to the United States with dengue infections every year, the virus may well succeed West Nile as North America's next viral invader.

But Rift Valley fever just might give dengue a run for its money. This ancient African denizen has the capacity to sicken and kill more people than a low-achieving invader like West Nile and is a fearful trade disrupter as well. Ferried by mosquitoes, RVF can lie dormant in mosquito eggs for years until a heavy rainfall sparks an outbreak.

The virus caught the attention of Europeans living in East Africa when it ravaged newly introduced livestock populations in 1930, causing abortions and high fatalities. At the time it also gave humans a mild flulike fever that could cause blindness. Two decades later, American biowarriors praised the virus's quick ability to cause 42-degree-Celsius (107-degree-Fahrenheit) fevers and rupture blood vessels in the eyes and concluded that it would be a "great demoralizer" if released among the public. In 1977 a hemorrhagic form abruptly appeared in Egypt, near the Aswan Dam, where it killed millions of animals, infected 200,000 humans, and emptied 700 people of their bodily fluids, killing them.

In Kenya RVF continued to prove its climate sensitivity by erupting every 10 years, after heavy rainfalls directly related to El Niño's weather dance in the region. After one of the worst downpours on record forced herders to crowd their livestock onto islands of dry land in 1997, the fever exploded with such deadly force that medical doctors at first suspected Ebola. Hundreds of Kenyans died with inflamed brains, distorted livers, and blood flowing from their gums, eyes, anuses, and ears. All in all, more than 80,000 people caught the disease while tens of thousands of sheep, goats, and camels vomited blood and perished. Veterinarians warned at the time that the epidemic "could expand considerably from its present focus" and spread across the Red Sea.

In 2000 the fever did just that. RVF leapt out of Africa into secretive Saudi Arabia and Yemen, where traders refused to report the disease until it had killed more than 10,000 animals and nearly 40 people. The infection probably arrived on sheep (Saudi Arabia is the world's leading sheep importer) and appears to have taken up residence in Jizan, a Red Sea port known for its heat and mosquito-friendly humidity.

Given climate variations, the accelerating livestock trade, and restless rounds of globetrotters, RVF is bound to win global citizenship sooner rather than later. Europe has a number of mosquito species that are just as fond of livestock and human blood as Saudi Arabia's. A 2004 report on vector-borne diseases on that continent even concluded, "Rift Valley fever must be considered a serious threat.… Should RVF virus be introduced in an infected animal from Africa or the Middle East, local transmission could occur." U.S. scientists arrived at similar conclusions years ago. Rift Valley fever just might deliver misery on an African scale to the backyards of North America.

So climate change has become another player in the dark game of species juggling. Although scientists continue to argue about how much microbial mayhem is actually being driven by climate, the point is probably moot. Climate is another disease trigger in an arsenal already overstocked by urbanization, travel, and trade. The emerging patterns aren't simple or pretty. Warm-weather parasites are eating all kinds of species, including trees and coral, at fantastic rates. Ticks and mosquitoes are having a party. Tropical bugs are moving to temperate cities where climate change helps provide a friendly welcome. Carbon dioxide and its warming effects seem to smile on creatures that can suck the life out of us over time.

Like the global meat trade and farming monocultures, climate change has introduced levels of instability and menace to the world that simply weren't there before. Disease invaders driven by floods or drought probably won't overwhelm us in North America or Europe, but they have already taxed our public health infrastructure and may well whittle our health away like fleas on dogs. It is also clear that climate riders are simplifying our living environments by reducing the number of plants and animals in our household.

A growing army of politicians and bureaucrats, who proclaim an undying faith in the power of technology to thwart tragedy, believe that humans can adapt fast enough to outrun climate change. But these techno-geeks ignore two facts. The first is that global change has already created several inflammatory alliances of people, parasites, plants, and animals that cannot be broken now. The glaciers are melting, the forests are on fire, and insects have packed a load of microbial invaders in their suitcases. The second is the historical record on climate. It shows that most civilizations haven't handled sudden changes in the weather any better than we have. Long-term thinking just isn't one of our stronger traits as a species.

For the last 10,000 years, climate has made and unmade civilizations with an impartial furor. Climate has humbled or hanged our elites time and time again. About 4,000 years ago, several Bronze Age cultures packed up and died because of unannounced warming that brought drought and crop diseases. Abrupt climate change rearranged the greatness that was Egypt with miserable cold and little rainfall beginning in 2150 BC. The bad weather produced this terrifying lamentation:

Lo, the desert claims the land.
Towns are ravaged.
Upper Egypt became a wasteland.
Lo, everyone's hair has fallen out.
Lo, great and small say, "I wish I were dead."

Lo, children of the nobles are dashed against walls.
Infants are put on high ground.
Food is lacking.
Wearers of fine linen are beaten with sticks.

In 540 BC another general crisis hit the planet when climate change (probably driven by volcanic activity) plunged most of the known world into a fearsome hell of extreme events. Bad weather in East Africa stirred up rodents and the plague, which trade ships promptly delivered to Rome and Constantinople. It eventually killed 40 million people. In the 14th century, climate drivers again woke rodents and more plague on the Asian steppes and gave Death another extended vacation in Europe.

The Maya erected such a complicated urban society that they even had a word for a 400-year time period. But in the ninth century, just 1,200 years ago, the driest weather seen in 7,000 years descended on the region for a decade. The people executed their rulers and abandoned their cities.

The Little Ice Age, between the 14th and 19th centuries, again illustrates that a few changes in temperature can set all sorts of black ships sailing. In the 1560s, a cold snap ended a long spell of mild weather in Sweden and set glaciers on a growth spree. The change put much of the country under water or in a deep freeze. Lowering the temperature stirred up an ugly mix of human, animal, and plant diseases as pointedly as elevated temperatures are now doing. One church parish register in western Sweden recorded this abrupt weather shift with a rude and graphic clarity. The Protestant declaration, like the Egyptian lament, remains one of the world's boldest testaments to how changes of a few degrees can unravel everything:

Because of our sins the weather deprived us of God's gifts both on
land and in the water. The clothes rotted on the backs of the poor. In
the winter the cattle fell ill from the rotten hay and straw which was

taken out of the water so that he who had a hundred goats and sheep
could not keep two, in fact not one of them. It went the same way
with the cows and calves and the dogs which ate their dead bodies
also died. The soil was sick for three years.... People ground and
chopped many unsuitable things into bread such as mash, chaff,
bark, bud, nettles, leaves.... This made people so weak and their
bodies so swollen that innumerable people died. Many widows, too,
were found dead on the ground with red hummock grass, seeds from
the fields, and other kinds of grass in their mouths. People were found
dead in the houses, under barns, in ovens of bathhouses and wher-
ever they had been able to squeeze in. Children starved to death at
their mothers' breasts, for they had nothing to give them to suck....
Gallows were built in our district [for thieves] and hung full.

Brian Fagan, a British anthropologist at the University of California,
Santa Barbara, doesn't think 6.5 billion people, largely concentrated in
cities and highly dependent on expensive fossil fuels, can adapt any more
gracefully to climate change than the ancient Egyptians or 16th-century
Swedes. He fears that we have traded up in vulnerability in order to
handle small troubles like West Nile or the odd drought or flood. But
we just aren't prepared for successive hurricane visits, rising sea waters,
or 40-year dry spells and all the other discord that global warming
promises.

Fagan compares modern technological society to a supertanker
piloted by inattentive fools:

Only a tiny fraction of the people on board are engaged with tending
the engines. The rest are buying and selling goods among themselves,
entertaining each other or studying the sky or the hydrodynamics of
the hull. Those on the bridge have no charts or weather forecasts and
cannot even agree that they are needed; indeed the most powerful
among them subscribe to a theory that says storms don't exist or if they

do, their effects are entirely benign, and the steepening swells and fleeing albatrosses can only be taken as a sign of divine favor. Few of those in command believe the gathering clouds have any relation to their fate or are concerned that there are lifeboats for only one in ten passengers.

Supertankers have always had difficulty changing course.

NINE

Nemesis: The Global Hospital

*"The medical establishment has
become a major threat to health."*
—Ivan Illich

S ometime in the fall of 2003 in the busy markets of Guangzhou, China, a place where cages of snakes, ferrets, and ducks all squirm for the attention of gourmands, a virus carried by an edible bat took up residence in a bunch of edible cats. (To the city's wealthy elite, the mongoose-like palm civet cat is a fashionable treat.) In the energetic cacophony of 10 million people, an ailing cat sneezed on a few animal traders or cooks. Restaurateurs then passed the virus on to customers innocently spooning a steaming bowl of dragon tiger phoenix soup flavored with chrysanthemum petals. And that's probably how a previously unknown coronavirus, soon to be dubbed severe acute respiratory syndrome (SARS), became another member of the emerging diseases club.

At first the Chinese didn't pay all that much attention to the new viral immigrant. It behaved like flu and killed like pneumonia, a common curse of sooty Chinese cities. The hardy and often scrofulous residents of Guangzhou simply dubbed the disease *fei dian xing fei yan,* "not the usual kind of pneumonia." They advised anxious foreigners to burn vinegar in their homes, get a shot (injections are the favored solution to every medical problem in China), and avoid overcrowded hospitals where doctors and nurses were rumored to be dying in heaps. When worried citizens started to hoard traditional medicines and ask a few questions, the Chinese politburo went to work and covered up the outbreak in the name of economic stability. The media, a reliable stabilizer, kept quiet too. The cold-like virus and intermittent lung destroyer might then have burned itself out in Guanghzou but for a series of unfortunate events that delivered the invader to hospitals abroad. That's where SARS exploded, resulting in one of the world's shortest biological invasions but possibly the most expensive: The bill hit $50 billion.

The explosion was no accident. Every day hospitals infect 1 in 10 patients with often costly or fatal infections by means of unwashed hands and general poor hygiene. Hospitals are full of sick people, after all, and the immune systems of the unwell don't work any better than those of a factory-bred, highly medicated chicken. In fact, SARS found everything in a hospital that a globally ambitious invader needs: lots of traffic, highly mobile staff, unclean quarters, massive overcrowding, and plenty of immunocompromised individuals. Walking a respiratory bug through the world's crowded emergency wards is like dropping a zebra mussel in the Great Lakes or introducing bird flu to a chicken factory.

SARS, however, differed in two respects from most of the pneumonia makers and blood poisoners that routinely invade hospitals. First, it didn't just overwhelm the unwell but targeted health-care workers. Most invaders, such as methicillin-resistant *Staphylococcus aureus* (MRSA), just afflict the afflicted. The viral invader also reversed the flow of hospital invasions. In most cases a crafty hospital invader erupts in an intensive care unit and then, over time, hops from patient to patient, enters acute care, and finally migrates to a nursing home or the greater community with little fanfare or reporting. SARS simply walked in the front door and illuminated the sorry state of infection control.

The grotesque and chilling odyssey of SARS began with a shrimp salesman. Sometime in January 2003, a 37-year-old seller of factory-raised shrimp picked up the virus, started hacking, and sought relief at Zonshang Traditional Hospital. There the unfortunate shrimp guy—the "poison king," as the Chinese now call the dead man—splattered mucus all over the place, infecting 130 hospital workers and patients. Liu Janlum, a 64-year-old professor of nephrology, picked up the bug while attending to other patients at the hospital.

The professor then took a flight to Hong Kong, 150 kilometers (95 miles) away, to attend a wedding reception. While staying on the ninth floor of the Metropole Hotel, he started to cough and feel feverish. Like the shrimp guy, Liu turned out to be a superspreader, an individual who can shower others with infection the way a cloudburst soaks a house. At the hotel Liu quietly rained the invader onto at least 12 fellow guests. Within days, three of Liu's hotel neighbors took the virus to Singapore's hospitals. Another two guests ferried the bat virus to Toronto's hospitals. And a 26-year-old airport worker who had visited a friend at the Metropole admitted the killer to Hong Kong's Prince of Wales Hospital.

In the emergency department the feverish airport worker effortlessly broadcast his viral news, and 11 hospital workers soon called in sick. In order to unclog the man's mucus-filled lungs, the doctors used ventilators and nebulized bronchodilators, which inadvertently dispersed the virus in tiny airborne particles. With beds just 1 meter (3 feet) apart in general wards, the invader spread from patient to patient. Within short order nearly 8 percent of the hospital staff and even the hospital's director were gasping for air, quarantined, bedridden, or in shock.

Peter Cameron, an Australian doctor who worked in the hospital's emergency ward (it was eventually closed to help contain the mayhem), later wrote that he had never seen a viral invader smite a hospital: "For fear of creating pandemonium, no moves were made to educate the public about preventive measures.... I remain extremely frustrated that others are not learning the lessons that we have learnt regarding the need for stringent infection control."

Another SARS superspreader took the invader by the hand to Singapore's 1,200-bed Tan Tock Seng Hospital. Crowded emergency departments and highly mobile medical staff did the honors with more help from superspreaders. One patient alone infected 24 health-care workers, 15 patients, and 12 visitors. Within eight weeks, peripatetic staff and patients spread the virus to five city hospitals and two nursing centers in a city of 6 million. Authorities got a handle on the epidemic only by closing the Tan Tock Seng, isolating the infected, donning masks, and exercising tight infection control. In the end Singapore recorded 206 SARS cases with 32 deaths. The majority were health-care workers or elderly patients.

The same wretched circus had a run in Taiwan. A 58-year-old traveler with a fever introduced the invader to Municipal Hoping Hospital in Taipei. It quickly found a superspreader in a dedicated hospital laundry attendant who ignored his diarrhea and pneumonia and doggedly stuck to his duties. The workaholic generated 137 infections among patients, doctors, and nurses. More than 10,000 of their contacts were placed in home quarantine. Several Taiwanese doctors later admitted (in the disarming babble of their profession) the awful truth: SARS "was transmitted rapidly within hospitals, which re-created pools of persons who became infected and through whom the disease was spread. Intrahospital transmission amplified regional outbreaks and augmented spread of illness into the community."

The outbreak in Toronto proved that North American hospitals were just as adept at broadcasting microbial invaders as their Asian counterparts. The 44-year-old son of a 78-year-old Hong Kong jet-setter developed fever and brought the invader to Scarborough General Hospital in late February. Before he died, the patient spread the virus to two roommates and a nurse. The infection soon ballooned into 128 cases among health-care workers, patients, and their visitors. One exasperated doctor revealed in the *Journal of Emergency Medicine* that "the presence of one unrecognized SARS patient waiting for a ward bed in one crowded emergency department in Toronto created the epicenter of a city's outbreak." Another patient infected seven hospital visitors, six hospital staff, two patients, two paramedics, a firefighter, and a housekeeper.

Toronto's first SARS outbreak in February and March was quickly followed by another cluster of cases in April and May fostered by more careless infection control, overcrowding, and chronic staff shortages. At the end of the virus's unscheduled 2003 visit, four Toronto hospitals largely accounted for 247 cases, 44 deaths, and the controversial quarantine of 23,000 people. Seventy-seven percent of the infected were patients, health-care workers, or hospital visitors.

Not surprisingly, Hanoi and Beijing also discovered the power of hospitals to seed infection and called it "the main characteristic of the SARS epidemic." A 2004 conference on the global invader pointedly described health-care workers as "bridges" from hospitals that carried the infection deep into many Asian communities.

Although the media still portray SARS as an exotic emerging microbe, the medical community soberly recognized the truth: SARS was just another honest-to-God hospital-made, or nosocomial, pestilence. In a 2003 edition of the *Singapore Medical Journal*, Paul Tambyah, an infectious disease expert, angrily blamed professional negligence for "the devastation caused by a highly transmissible nosocomial viral infection.... One can only hope that we will never return to the era when we treated respiratory isolation in a cavalier fashion, when we ignored hand washing guidelines, 'forgot' to gown and mask for procedures, and insisted on keeping patients crowded into beds near enough for efficient nosocomial transmission of large droplet nuclei."

A 2004 Canadian report *(SARS and Public Health in Ontario)* also declared the ugly truth: "SARS was largely a hospital spread infection...."

What is clear from SARS is that hospitals can become epicenters of infectious outbreaks that can move into the community." One prominent Toronto physician added another pertinent observation: "In our drive to technology in the 1980s and 1990s, we forgot the basics."

The basics include not only good hand-washing but also a simple respect for elemental biological truths. Hospitals are hazardous places because they house monocultures of overmedicated sick people with battered immune systems. These deliberate collections of the feeble, the infected, and the near dead attract armies of bacteria, viruses, and fungi (including most drug-resistant types) for the same reason that genetically uniform fields of wheat or coffee entertain crowds of micropredators. The indiscriminate use of antibiotics to fight the microbes has in turn created fiercer and more resistant hospital invaders, just as pesticides have selected for the nastiest and toughest crop molesters.

A 1960 medical tract on nosocomial infections (even then "a major public health problem of major importance") summed up the problem: "Congregating a large number of sick people under a single roof has many advantages but one serious drawback—that the disease from which one is suffering may be transmitted to others." The modern hospital, a kind of mall for sick people, has compounded that reality by treating more immunocompromised and elderly individuals than ever before with more and more invasive procedures from cataract surgery to knee replacements.

The scale of biological invasions unsettling hospitals now dwarfs the whole SARS debacle but gets no respect. Sir Alexander Ogston, a 19th-century Scottish surgeon, was correct when he observed that "human nature forgets unseen foes." In the United States, prominent invaders such as methicillin-resistant *Staphylococcus aureus* (MRSA) and vancomycin-resistant enterococcus (VRE) sicken up to 2 million patients a year and kill 103,000. Dirty hospital hygiene combined with unprecedented medical poking and prodding slaughters more Americans than AIDS, auto accidents, and breast cancer combined.

The invaders are hardy and diverse. *Clostridium difficile,* a nasty bacterial colon invader, can survive up to five months on a hospital floor. Hepatitis B virus can contaminate electroencephalographic electrodes for up to seven days. *E. coli,* which accounts for 12 percent of all blood poisoning in hospitals, does very well on plastic surfaces. *Klebsiella pneumonia,* an opportunist that can invade the lungs, urinary tract, and abdomen, kills more people every year than SARS and travels on mobile patients. *Salmonella enteritidis,* a classic immune system buster, can arrive in hospital food. A group of fungal invaders, members of the *Candida* genus, are now the fourth most common bloodstream pathogen in hospitals.

The problem is inescapably global in scale. Every year approximately 5 percent of the world's 6 billion denizens visit a hospital, the icon of scientific progress. As a consequence the medical fraternity treats about 300 million sick souls a year. Current hospital-acquired infection rates hovering between 5 and 10 percent (an extremely conservative estimate) mean that hospitals give at least 15 million individuals a viral, bacterial, or fungal infection they didn't arrive with. Of the 15 million who annually battle unwanted infections, approximately 1.5 million die from them. According to WHO statistics, that's nearly triple the annual death toll from influenza or almost equal to the annual death rate for tuberculosis: 1.7 million. "Thus hospitals in some areas of the world may be playing a role as amplifiers of disease, rather than as consolidated institutions designed to relieve or at least palliate some diseases," observed a Mexican hospital infection specialist in 2003.

Some of the most appalling mortality statistics come from so-called developed or healthy countries. In Canada, a nation with 30 million citizens, hospital infections caused by unwashed hands, contaminated equipment, or overcrowding afflict as many as 200,000 patients and send 10,000 citizens to the morgue every year. Britain's notoriously dirty hospitals infect approximately 300,000 patients a year and probably bury at least 20,000. In California, a sunny bastion of high-tech progress, hospital infections kill or cripple more citizens than car accidents. And in Pennsylvania, the first American state to collect really good data on hospital-acquired infections, which it began doing in 2004, invasions sicken 12,000 and dispatch 1,500 patients a year at a cost of $2 billion.

Hospital invaders are just as expensive as crop or livestock barbarians. According to the U.S. Committee to Reduce Infection Deaths, these largely preventable invasions cost American patients about $30 billion a year. Not only do invaded patients cost nearly three times more to treat with drugs than their non-infected peers, but they also stay in hospital longer or require more surgery. A urinary tract infection, for example, boosts hospital costs by 47 percent; a pneumonia infection introduced by an unclean respirator adds $40,000 to the total bill; and a bloodstream infection caused by a catheter colonized by MRSA might bring costs up by $30,000. So if hospital infections don't kill, maim, or cripple us, they might eventually bankrupt the health-care system altogether.

Nearly 30 years ago the Catholic philosopher Ivan Illich warned about something he called "medical nemesis." He argued that irreparable damage would follow the industrial expansion of medicine, just as irreparable damage has dogged the industrial expansion of livestock operations. Whenever professionals seek to banish all pain, sickness, and even death itself with more drugs and surgeries, they ultimately create their own health-destroying enterprise. Medical progress, warned Illich, "has come with a vengeance which cannot be called a price."

Illich, who died with dignity and without a doctor in the room, regarded hospital-acquired infections as just one part of medical nemesis. He argued that whenever *Homo economicus* turned into *Homo religiosus,* the backlash would be severe. He measured that severity not just in the rate of hospital-acquired infections but also in unnecessary surgeries, wrongly prescribed drugs, and the drive to prolong life mechanically at any cost.

The mind-boggling scale of medical nemesis was recently revealed in a study in the *International Journal of Sexually Transmitted Diseases and AIDS*. In 2004 three doctors from the World Health Organization calculated how many cases of hepatitis B, hepatitis C, and HIV had been caused directly by the reuse of hypodermic needles. For the year 2000 alone, they found that contaminated injections by physicians had caused 21 million hepatitis B infections (a formidable man-made invasion), 2 million hepatitis C infections, and a quarter of a million HIV infections. Their blunt conclusion: "Injection overuse and unsafe practices account for a substantial burden of death and disability worldwide."

No invasion tale illustrates medical nemesis and the sordid state of hospital-acquired infections better than the remarkable progress of methicillin-resistant *Staphylococcus aureus* (MRSA). It's infiltrated most North American hospitals and now accounts for more than 50 percent of all staph infections in hospitals worldwide. For the first time, British patients as well as the Consumers Union in the United States are demanding better hospital hygiene and full disclosure of infection rates. The ambitious superbug has also started to spill out of hospital wards and nursing homes into sports clubs, jails, military bases, schools, and day-care centers. What doctors once regarded as a source of treatable wound infections confined to surgical wards seems poised to become another drug-resistant global invader. Think of it as a bacterial version of avian flu in slow motion.

Staphylococcus aureus, the most common hospital invader, is an ancient bacterium that can live on any surface and is almost as hardy as anthrax. It has long colonized the human body; about 30 percent of any given human population carry the germ in their nostrils or on their skin. The bacteria remain benign and even useful occupants until the immune system trips and falters. Whenever the body's defenses fail, the waiting barbarian starts to go to work under a number of guises. It can appear as a skin or soft-tissue infection (a boil or an abscess). But once it breaks through the skin, it can take the form of pneumonia or blood poisoning or even heart inflammation. No pathogen can create as many diseases as staph, perhaps the best-armed and most versatile bacterium on the planet. It is the celebrated author of toxic shock syndrome, too.

MRSA obeys the invader's code with remarkable precision: Just a few can make a hell of a difference, and fast. It can also employ the help of other invaders, including a single-cell amoeba. Once inside the amoeba, MRSA can increase its numbers a thousandfold and produce highly invasive offspring that are more resistant to biocides and antimicrobial cleaners than relatives born freely. The organism can colonize patients and medical personnel without producing symptoms for months on end. A carrier or superspreader can saturate a hospital in short order by contaminating medical equipment, staff clothing, room surfaces, and even bathtubs. The bug can occupy dirty cabinets, tables, bed rails, doorknobs, or toilet seats. It can travel on ties, coats, shoes, and uniforms from patient to patient. More than 60 percent of the surface area of a burn unit room can be contaminated with MRSA. Whenever health-care workers lean over a sick person infected by MRSA, 60 percent typically pick up the superbug on their clothing. MRSA can also colonize stethoscopes and blood pressure monitors and survive on a computer keyboard for 24 hours. Even masked surgeons with MRSA in the nose can pass on the bug to elderly patients during, say, hip replacement procedures—and kill them. But one of MRSA's best routes remains overcrowding. Repeated hospital studies have shown that whenever the distance between beds decreases from 2.5 to 1.9 meters (8 to 6 feet), the spread of MRSA among patients increases threefold. "Once a strain colonizes a hospital, it's like bagging a cat," admits Donald Low, a microbiologist at Mount Sinai Hospital in Toronto. "You get an organism in one corner and then, bang, you've got one across the room. It's like holding Jell-O in your hands."

S. aureus has lurked in hospitals for hundreds of years. It typically took advantage of surgical wounds to explode and once accounted for a third of surgical deaths. But during World War II it started to make some important moves. First it aggressively displaced another invasive cousin, streptococcus, as the most common bacterial invader in hospitals. Then it diligently acquired resistance to penicillin by learning how to make an enzyme that simply ate the drug for dinner. Whenever another new antibiotic threatened its hospital conquests, the bacterium mutated or searched among its bacterial cousins for the genetic information necessary to render those drugs harmless too.

In 1960 a few naive scientists thought they had finally outwitted staph with the introduction of a synthetic version of penicillin called methicillin. But the bug put out a distress call on a bacterial intelligence network and quickly found some resistance tools. Aid in the form of a drug-resistant gene was found, amazingly, on a bacterial relative living on the skin of squirrels. The new gene gave *S. aureus* the ability not only to defeat methicillin but also to withstand an army of other antibiotics. The American journalist Michael Shnayerson notes in his book *The Killers Within* that MRSA is entirely a man-made creation: "It wouldn't exist without the drug that spawned it." Had hospitals not employed methicillin, *Staph aureus* wouldn't have been forced to seek superbug powers.

In 1963 the first outbreak of MRSA swept through 48 wards of Queen Mary's Hospital for Children in Surrey, England, where it attacked 37 children and killed 1. The bug quickly went global, making appearances in Turkey, Australia, and India, where it rendered many hospital infections untreatable. At a 1970 conference on nosocomial infections, one doctor reported that this particular MRSA pandemic "certainly awoke many to the deficiencies in our routines for the hygienic care of patients in hospital."

Physicians in Denmark and Holland, however, decided to bury these deficiencies rather than bury patients. After losing far too many patients to MRSA, both the Danes and the Dutch adopted a "search and destroy" policy in the 1980s. First, they screened incoming patients and high-risk surgical candidates for MRSA and treated them in single rooms. They also routinely monitored the bacterial carriers on staff. Those found to carry the superbug were sent home, treated with antibiotics, or even ordered to have their tonsils removed. Whenever a patient was invaded, doctors closed the ward and tested all patients and staff for colonization or infection. Colonized doctors and nurses were sent home until they got rid of the invader. In addition, everyone diligently washed their hands after treating patients. Studies show that whenever hospital staff improve their hand-washing routines by 35 percent, the rate of MRSA infections plummets by half. All in all, these cost-effective interventions gave Danish and Dutch hospitals the lowest proportion of bloodstream infections caused by MRSA in the world: 1 percent. Although a handful of American hospitals have obtained the same results by employing the same tools, most accept horrifically high rates of MRSA. According to the U.S. Committee to Reduce Infection Deaths, just 2 percent of all hospital-acquired staph infections were MRSA in 1974; by 2003 that figure stood at 57 percent.

Like the United States, Britain took a laissez-faire approach to the invader and allowed MRSA to explode under medical noses. As a consequence, MRSA cases increased by 600 percent in the 1990s and are expected to double again in the next 10 years. Britain's hellishly persistent epidemic has killed thousands of citizens of all ages (including the mother-in-law of the former Conservative leader Michael Howard) and has sorely diminished the reputation of the National Health Service (NHS). Microbiologists have sourly noted that even Britain's slaughter-houses and highways are a darn sight tidier than its hospitals. "Where MRSA is concerned, Britain has become the sick man of Europe," recently wrote Hugh Pennington, an erudite Scot and president of the Society for General Microbiology.

The British MRSA crisis has created such an uproar among indignant citizens that in 2003 the government forced hospitals to post their MRSA infection rates on the front door, so patients and their relatives know when they are walking into potential trouble. When that long overdue measure (Florence Nightingale advocated for similar transparency) had little or no effect on MRSA levels, the government threatened to fire hospital directors who ignored the MRSA threat. "People go into hospital to get better," complained one Liberal Democrat MP to the BBC. "But they are getting sicker because infection control is not a high priority."

Penny-pinching ideologues and careless professionals set off the invasion in the 1980s as wave after wave of MRSA strains flooded hospitals in England and Wales. In 1992 a particularly nasty strain (EMRSA 16) infected 400 patients and 27 health-care workers in a general hospital, a geriatric facility, and a mental hospital. Two years later it jumped into another 21 hospitals and entered London's sprawling institutions. Transferred patients carried it from one sick building to another. By 2000 MRSA had thoroughly colonized British hospitals, raising the proportion of bloodstream infections caused by the superbug from 2 to 40 percent. Only small rural hospitals escaped the bacterial conquest. The number of death certificates actually listing MRSA (something doctors rarely do with hospital-acquired infections) climbed from 51 in 1993 to 800 in 2002.

The superbug, of course, had lots of determined administrative assistance. Britain happens to have the most crowded (and thereby most unhygienic) hospital system in Europe. Ever since the NHS contracted out cleaning services to multinationals in the 1980s, cleanliness has become a hot issue in British hospitals. It's no secret that the nation's dirtiest hospitals are mopped up by private contract cleaners. The bug also exploited fiscal reforms that assumed that shuttling patients around wouldn't have any microbial consequences. After severe reductions in beds for acute and chronically ill patients, the health service started to transport patients to where spaces were available, a practice now known as "hot bedding." Old folks colonized by MRSA were sent back to nursing homes, where the bug colonized more people, only to be readmitted to acute care hospitals a few days later, where MRSA took advantage of any available human transport or mistake. Fiscally motivated cuts in isolation wards and cleaning staff also fed the epidemic. As a result the percentage of staphylococcal bloodstream infections caused by the superbug rose exponentially in children, from 1 to 13 percent. Doctors, by the way, describe MRSA blood poisoning as "a significant risk factor for death."

By the late 1990s, the situation in British hospitals had gotten very grim. In Edinburgh an outbreak closed down the main heart unit. Alarming cases of vancomycin-resistant *S. aureus* also appeared. Vancomycin is the last antibiotic still capable of battling MRSA. Damning BBC investigations about superbugs and lawsuits from angry spouses who lost partners to the infection soon followed. So, too, did a 1998 House of Lords study on shoddy infection control.

But the well-defined Danish-like solutions were ignored, as was the advice of microbiologists. As early as 1959, several microbiologists had warned that antibiotic-resistant staph infections required active cooperation, extra work, and the isolation of carriers. To this day many disease experts believe that a simple reduction in overcrowding combined with some fanatical hand-washing and isolation of carriers would have stopped the bacterial invader cold. Nevertheless hospital administrators and politicians continued to support the medical equivalent of globalization: treat as many far-flung patients as possible, as quickly as possible.

The scale of the epidemic and its human costs didn't really become evident until the United Kingdom's National Audit Office (NAO) began to spell things out in 2000. Written in the deadliest of technocratic prose, it highlighted a disturbing trend: In 1995, 189 hospitals in England and Wales reported MRSA outbreaks that infected three or more patients. Three years later the nation's hospitals logged 1,597 incidents. These invasions probably cost the health system nearly £1 billion ($1.6 billion) a year. Patients colonized by MRSA typically remained in the hospital two and a half times longer (an average of 11 extra days) and cost the system nearly three times more to treat than uninfected patients. The NAO recommended better hand-washing, monitoring, and reporting.

Three years later the NAO auditor revisited the MRSA controversy and found few solid improvements: "The increased throughput of patients to meet performance targets has resulted in considerable pressure towards higher bed occupancy which is not always consistent with good infection control." In other words, administrators continued to juggle patients and their microbes about in order to fill all the beds and in so doing served as the dedicated connectors for MRSA colonization. Not surprisingly, the number of blood poisonings caused by staph infections had jumped by 8 percent, from 17,933 in 2001 to 19,311 in 2003. Moreover, the proportion of bloodstream infections caused by MRSA in Britain remained "the worst rate in Europe" at 44 percent. "There continues to be non-compliance with good infection control practices," added the report.

Claire Rayner, a former NHS nurse, learned about the sorry state of compliance firsthand. In 2004 she picked up an MRSA infection while recovering from her fourth knee operation in an orthopedic hospital. Even after the institution posted a "Take Care, MRSA" sign on her room, little care was shown: "I was shocked to see that the nurses still didn't wash their hands after they looked at me. And the dirt around the hospital was horrifying. Bedpans were left at the side of the bed for God knows how long.... In the past few years, I've spent more time in hospitals than I care to remember. After what I've seen, I would have to be seriously ill, in fact, at death's door, before I would go and stay in another."

Hardly a week now goes by without a British newspaper publishing one MRSA horror story after another. Many end up being posted on the MRSA Watch Web site or on British Nursing News Online. The stories make lurid and depressing reading. A 53-year-old woman goes in for a routine ankle operation and dies of an MRSA blood infection. Hospital infections kill more Scots every year than drug overdoses, emphysema, and AIDS. A snap government inspection finds London-area hospitals "unacceptably dirty." A healthy 87-year-old goes in for hip surgery and an MRSA invasion keeps her there for five months battling for her life. A *Sunday Times* crew takes swabs around beds and showers in a local hospital and discovers that a busy urban road would be a safer place to get ill. A patient with pus flowing from his MRSA-infected wound shares the toilet with other patients and no one notices. A 29-year-old woman goes in for surgery expecting a five-day stay but ends up spending four weeks fighting MRSA. ("For days I felt like a vegetable," she reported.) The 45-year-old Scottish pop star Edwyn Collins survives brain surgery but his MRSA proves "very difficult to get under control." A prudent doctor discloses that he routinely discharges patients early because his hospital "isn't a terribly safe place to be." A girl survives the London bombing of 2005 only to nearly succumb to MRSA-infected wounds in her foot. A 64-year-old enters a hospital for a routine hernia and four months later leaves in a coffin because of MRSA and another common bacterial invader.

In 2004 Edward Leigh, MP and chairman of the parliamentary Committee of Public Accounts, assessed the carnage and publicly rebuked the health system for its inaction and ineffectiveness: "It is astonishing that poor ward cleanliness, lax hand-washing practices, a shortage of isolation facilities and high bed occupancy rates are still plaguing NHS hospitals." Leigh also identified another problem: The government still had no national surveillance and reporting scheme for hospital invaders other than MRSA, which accounts for less than 10 percent of the pandemonium in hospitals.

MRSA is now so entrenched in British society that it can be found pretty well everywhere. Birds carry the drug-resistant organism and so do dogs and cats. Veterinary health-care workers serve as reservoirs, as do many doctors and nurses. Chris Malyszewicz, an independent chemist, randomly tested 10 people on the streets of Northampton and found that 7 carried MRSA. He also found the invader on dirty banknotes in London. "How bad is it going to get before the Government do something?" he asked *The Northampton News* in 2005.

Despite concerted campaigns to wash hands and isolate patients, half of Britain's hospitals will fail to meet public targets of halving their MRSA rates by 2008. And so the killing and affliction continues. One week MRSA kills a two-day-old baby and the next it colonizes Ronnie Biggs, one of the masterminds of the Great Train Robbery. Even former hospital cleaners have been felled by the bug. When a 42-year-old former NHS cleaning supervisor went in for routine surgery, he ended up losing an entire leg to a rapid and explosive case of MRSA: "I have been disabled for the rest of my life ... all because of MRSA."

But poor Britain isn't alone. More than 50 percent of the staph infections raging through Colombian hospitals are MRSA. In Thailand one 2006 hospital study found that MRSA infections killed 50 percent of the children they infected, leading researchers to conclude that it "may be responsible for a considerable burden of disease in Asia." Canada, the Great White North, has recorded a tenfold increase by the invader in the last decade. A 2003 report by the Canadian infection control expert Dick Zoutman tells a familiar story. He found that 40 percent of 172 hospitals in Canada responding to a survey didn't have adequate staff or protocols to monitor bacterial traffic. More than 50 percent didn't meet the minimum recommended level of one infection control professional per 250 beds. And only 13 percent of the hospitals actually did a proper job of tabulating the whereabouts of hospital invaders. Zoutman called the study a "wake-up call," but there's no indication that most Canadian professionals are listening any harder than their British peers.

Meanwhile MRSA has used American hospitals, where the bug accounts for more than half of all staph infections, as a stepping-stone to greater conquests. In the late 1990s, a variety of new strains started to attack aboriginal children in Minnesota, football players in Mississippi,

THE MRSA REVOLUTION IN EUROPE

The United Kingdom's crowded and dirty hospitals support
the highest rates of MRSA bloodstream infections in Europe.

Copyright © UK National Audit Office.

school kids in Chicago, and inmates in the crowded Los Angeles county jail. The new strains proved incredibly fit: They could turn boils into gross abscesses and melt lungs in 24 hours with devastating pneumonia. By 2005 MRSA had gone global. An 18-year-old Royal Marine died after simply scratching his leg on a bush, giving one of the new strains a portal into his body. It produced a deadly toxin (Panton-Valentine leukocidin) that wiped out his blood cells.

Community-acquired MRSA is now beginning to shake up the sports world, where skin contact and skin injuries are common. The disease has popped up in wrestlers, rugby players, and even canoers. Athletes with no history of chronic disease, immunosuppression, or antibiotic fatigue are now being afflicted by flesh-eating abscesses and Biblical-scale boils. Five members of a fencing team in Colorado got so ill that three required repeated hospital visits. In Pennsylvania and California seven football players ended up in the hospital. One required the removal of dead tissue and skin grafting. Crowded living conditions at training camps and the communal sharing of soap and towels probably spread the invader. The superbug also appears to have a fondness for Astroturf. Many coaches now say they would rather have their teams use 100 towels in practice "than come down with staph."

Family doctors are now seeing something they have not seen before: MRSA infections among normally healthy children. Since the late 1990s, doctors have witnessed a 25-fold increase in community-acquired MRSA. Two-thirds of the children with staph infections in the Chicago area now carry a strain of the superbug. It has caused devastating pneumonias and violent bone and skin infections. "We are admitting patients by the flocks with this," reports Dr. Robert Daum. "It may become a significant public health problem in the near future." It is also traveling fast: In 2006 the U.S. Centers for Disease Control calculated that as many as 2 million Americans now carry community-acquired MRSA. Three percent of American military recruits now carry the bug, and a whopping 38 percent of those carriers go on to develop soft skin abscesses or nasty necrotizing pneumonias. "The rapidity with which this has emerged over the last two to three years is probably unprecedented," Donald Low told *The Los Angeles Times* in 2006. "When you look at the numbers it way outstrips other so-called new infectious diseases."

M RSA's global spread is another reminder that hospitals have always been dangerous places. The historical record plainly shows that these institutions have launched an army of invasions ("hospital fevers" and "hospital gangrenes") as they've become bigger and more impersonal. But they didn't start out that way. In the beginning hospitals offered solace to the sick and dying on a small and homely scale. Ancient Arab physicians so appreciated the importance of cleanliness that they even built separate wards for the infected. Before building a hospital these observant doctors even hung up pieces of meat around a city in order to select the spot where the least putrefaction set in.

Throughout most of the Middle Ages, hospitals typically served as asylums for lepers and the poor. Run largely by monastic orders, these beneficent shelters offered "great spiritual solace but minimal physical comforts," says Guenter Risse, an eloquent hospital historian. By the 20th century, hospitals had grown so large and complex that they had reversed the priorities, with a cold focus "on individual physical reha-bilitation in more fragmented and depersonalized environments."

As hospitals evolved from merciful houses to big-box medical malls, doctors noticed that they were losing more and more patients to a mysterious tide of hospital-made maladies. In the 16th century, Ambroise Paré, the great Parisian surgeon, pointed out that hospitalized patients nourished more infected wounds than patients cared for at home. (At the time the infamous Hôtel-Dieu often squeezed as many as nine patients to a bed.)

Sir John Pringle, a British military doctor and hygiene reformer, observed in the 18th century that "the chief causes of sickness and death in an army" were "the Hospitals themselves." James Tilton, an American doctor, made the same damning observation during the American Revolutionary War, when 10 soldiers died of "camp diseases" for every one killed in battle. He argued that centralizing the sick in one institution just guaranteed that half of the army would eventually be "swallowed up" by disease.

In the 19th century, Sir James Simpson, a brilliant Scottish surgeon, fingered overcrowding and "hygiene evils" as the author of "surgical fever" (then a prominent invasion caused by *Streptococcus pyogenes*). In teeming urban hospitals with 400 to 800 beds, he observed, one in two amputees didn't survive surgery. Meanwhile, in small rural hospitals, only one in nine patients died. Simpson even coined a great word for this hospital-induced sickness: "hospitalism." But the profession didn't appreciate Simpson's clarity. "I have found that to most professional minds it seemed to be altogether a kind of medical heresy to doubt that our numerous and splendid hospitals for the sick poor could by any possibility be aught than institutions as beneficial in their practical results." Although he advocated for small cottagelike facilities, hospitals, like dumb dinosaurs, just got bigger.

Ignaz Semmelweis, a noble obstetrician, also recognized the profound harm hospitals could do. In the mid-19th century, puerperal fever, an infection caused by group A streptcoccus, was the MRSA of its day. It exploded when hospitals started to encroach on the maternity business and forced poor mothers to give birth in the institution instead of the safety of home. After the invader killed millions of women after childbirth, Semmelweis discovered why.

As director of Vienna's lying-in hospital, he noticed that new moms at one of his clinics died in large numbers while a neighboring maternity ward experienced few deaths. The critical difference proved to be doctor hygiene. Medical students who had just dissected cadavers regularly assisted in one clinic, while midwives, who left the dead alone, just delivered babies. Semmelweis connected the dots and boldly suggested that his profession wash up with chlorinated lime before examining pregnant women.

After he ordered a cleanup, mortality rates at Semmelweis's clinic plunged from 12 to 2 percent in a month. But one-fifth of mothers-to-be who attended birth clinics in Vienna and Prague continued to die from "doctor's plague." Fellow obstetricians didn't appreciate Semmelweis's angry advocacy for soap. When they didn't change their habits, Semmelweis compared a few of his obstinate peers to the fiddling Roman emperor Nero, which didn't go over well. In fact, many professionals adopted the very modern view that these hospital-caused fevers were inevitable and expressed determined indifference to the evidence for hand-washing. (Their modern counterparts still argue that it is okay to lose patients to invaders as long as you don't exceed the national average.) The profession's arrogance eventually drove poor Semmelweis mad. After being confined to an insane asylum in 1865, he died after a vicious beating by the staff gave him a fatal strep infection. The man who tried to end "hospitalisms" was, ironically, silenced by one.

The current state of hospital hygiene would still drive Semmelweis crazy. Since the advent of antibiotics, many institutions have let down their guard, grown bigger than feedlots, abandoned good barrier nursing (an older term for infection control), fired on-staff cleaning crews (it's all contract work), and let the germs in, two by two. According to medical journals, compliance with hand-washing rules rarely exceeds 40 percent at most hospitals because of carelessness, lack of sinks, poor training, hectic schedules, or lack in faith in hand-washing protocols. There are persistent gender differences too: More male physicians neglect basic hand-washing than females. So the patient-killing attitudes and habits that Semmelweis fought are still driving invasions today.

They have also played a starring role in the harrowing progress of *Clostridium difficile*. In just six years it has gone from being another "nuisance pathogen" for patients over age 65 to a deadly hospital invader that's rapidly eclipsing MRSA in mortalities. In 2003 it felled nearly twice the number of global SARS victims, and in just one urban Canadian hospital system, without making a single distressing headline at the time. The frightening epidemic, which is going global thanks to the profligate use of heartburn medications and antibiotics, clobbered patients, horrified their relatives, and simply overwhelmed hospital doctors. One chastened group of Montreal physicians later wrote that "C. diff" had shattered their confidence in battling infectious diseases altogether. The uncontrollable dying had also led to a "stark realization that our mastery of the microbial world is not absolute."

C. difficile started out as a rather harmless diarrhea-maker. The ubiquitous soil and livestock bacterium (its hardy cousins cause tetanus, botulism, and gangrene) first grabbed doctors' attention in the 1950s. That's when widespread antibiotic use in hospitals gave it unlimited opportunity to invade the guts of elderly or sick patients. Because most antibiotics tend to wipe out competing bacteria in the colon, *C. difficile* simply grew like mad.

In the 1970s, researchers learned that the bacterium had a few other tricks up its sleeve. It could produce two deadly toxins that could severely inflame the colon and cause a life-threatening degree of diarrhea. They called the new hospital-centered syndrome *Clostridium difficile*-associated diarrhea (CDAD). But for the most part doctors kept the bug under control by prescribing more and more powerful antibiotics. Experienced clinicians rarely saw more than three severe CDAD cases in a decade. "It was almost unheard of," recalls Sandra Dial, an intensive care physician at the Montreal Chest Institute.

But around 2000 the bug stealthily changed its behavior, and it soon exploded in a number of hospitals. One outbreak in Calgary in 2000 infected more than 1,000 patients; another one in Pittsburgh in 2001 killed 18 patients and so damaged the colons of another 26 patients that they all required colostomies. But because *C. difficile* wasn't an identifiable or reportable disease, these outbreaks went largely unnoticed.

Two years later a hypervirulent strain of *C. difficile,* observed in the Montreal area, started to produce 15 to 20 times more toxins. It exploited the same fertile terrain as MRSA—crowded hospitals full of elderly patients and with poor infection control—and spread like wildfire. The epidemic began by killing elderly patients in Sherbrooke, a city two hours' drive southeast of Montreal. It quickly raised the incidence of *C. difficile* infections among people over 65 tenfold, eventually bringing it to a harrowing rate of 866.5 per 100,000. It soon spread everywhere on the hands of doctors or the bodies of mobile hospital patients.

Sandra Dial was one of the first clinicians to notice a dramatic change in her patient caseload. In one month alone in the winter of 2002/2003, she saw five elderly immunocompromised patients with *C. difficile.* All were suffering from septic shock, extended bellies, continuous diarrhea, and severe inflammation of the colon. None responded to standard treatment. "We had never seen it before and didn't know what to do. They were getting sick and dying." At first Dial thought it was just one of those medical blips. While SARS distracted the country's attention, the *C. difficile* outbreak waned, and it seemed to disappear in the summer of 2003.

But it came back with a bang that winter, and the dying started all over again. The invader picked off the unwell and the well alike. It quickly spread from hospital to hospital, traveling with transferred patients and their infective diarrhea. Unwashed hands, dirty bathrooms, and harried staff all served as unwitting bacterial transporters. A healthy 50-year-old went in for elective surgery and died four days later from *C. difficile.* A retired allergist made the same mistake and nearly died. At that point, Dial made more inquiries and learned that the epidemic was largely a Montreal affair. After asking more questions, she put together some initial numbers. They were alarming: In just one 18-month period a so-called nuisance pathogen had infected

1,400 patients in six Montreal hospitals, killing 79 patients. She said to colleagues, "We have to do something and inform the public."

The big announcement didn't come from the hospitals or government but from the *Canadian Medical Association Journal*. In June 2004 it indignantly declared that *C. difficile* had killed more people than SARS, yet "neither public health nor hospital officials have warned the public." It added that the bug was even afflicting patients in their 20s and 30s with bloody diarrhea. Many required colostomies. One doctor and several other health-care workers caught the bug. Health-care workers also lost relatives to the epidemic. One nurse lamented that her healthy 80-year-old mother was admitted to a ward "full of people with *C. difficile*" and died five weeks later.

In a later report, four prominent doctors asked what was going on in Montreal and provided a familiar answer: "Our willingness to tolerate hospital rooms with four patients and a single bathroom, less than three feet [1 meter] between beds and progressively fewer resources assigned to housekeeping all facilitates the spread of this disease, as does our inability to achieve acceptable levels of hand hygiene among hospital staff."

Dial also added another driver: the increased use of proton pump inhibitors, a class of strong heartburn medications. Normal stomach acidity is a great bacteria neutralizer, but Dial suspects the overuse of acid suppression therapy in hospitals may have given *C. difficile* the upper hand. Initial studies suggest that the bug just produces more toxins in the presence of heartburn medication.

In a retrospective analysis of the epidemic in 2005, Jacques Pepin, a Montreal epidemiologist, assessed the damage. He calculated that *C. difficile* infected 14,000 hospital patients in the province of Quebec between 2003 and 2005, probably killing between 1,000 and 3,000 patients. His dismal report reads like many of Britain's MRSA documents: "The lack of investment in our hospital infrastructure over several decades, with shared bathrooms being the rule rather than the exception, may have facilitated the transmission of this spore-forming pathogen, which can survive on environmental surfaces for months." Not surprisingly, *C. difficile* is now overwhelming hospitals in the United States and Britain. The British caseload alone increased from 1,000 in 1990 to 43,672 in 2005.

Like MRSA, *C. difficile* is also scouting out new ground. In the United States it has invaded 31-year-old mothers and 10-year-old girls, populations normally regarded as low-risk. Swedish doctors have found that half of their cases have no immediate history of hospitalization. In Ireland more than 10 percent of the patients had not been in hospital for the 60 days before their illness. A review of *C. difficile* cases in England by Dial found that 70 percent of patients had not been admitted to hospital in the past year and that fewer than 50 percent had taken antibiotics in the three months prior to developing disease. "It is now evident that *C. difficile* disease has left the hospital setting to reach the community and that antibiotics are not the only culprit," noted a Montreal epidemiologist, Samy Suissa. So another horse has left the proverbial barn.

The solutions to crippling epidemics of MRSA and other invaders can be found in the work of 19th-century hospital reformers. Washing hands can save lives, and breast-feeding can protect infants from infections better than drugs. Smaller, houselike hospitals will always be safer than big boxes. Limiting antibiotic use is a wise counsel that needs to be actually practised. Treating elderly patients at home does less harm. Restricting the mall-like traffic inside hospitals might end a lot of microbial traffic. Adopting Danish and Dutch ideas of hygiene will save lives and money. Forcing hospitals to publish their astounding infection rates upholds the patient's right to know. And on it goes.

MRSA and *C. difficile* are just the initial shock troops of hospital invaders spilling into the greater community. An aging population combined with poor hygiene and growing antibiotic resistance (nearly 90 percent of bacterial invaders in Asia are resistant to front-line antibiotics) guarantees that hospitals will become more blighted than our oceans and livestock. So, too, does a sullen and persistent indifference to the urgency of the invasion. Hospital administrators and politicians seem to want to prevent biological invasions no more than shipowners want to clean up ballast water or than Big Beef wants to test its animals for BSE in North America.

In a 2004 editorial in the *Journal of Infection Control and Hospital Epidemiology,* Allison McGeer, a Toronto epidemiologist who both combated SARS and contracted the disease, openly criticized the apathy of her peers. She concluded that the major problem was "that neither healthcare professionals nor healthcare administrators are willing to invest intelligently in prevention." (The ecologist Daniel Simberloff has made the same comments about the United States Department of Agriculture and its lack of willingness to curb plant and animal invaders.)

In another issue of the same depressing infection journal, an American colleague, William Jarvis, took up McGeer's concerns and asked how long hospitals could tolerate their MRSA rates, given the ever-rising body count: "Why should patients have to accept the enormous and still expanding burden of excessive hospital stays, use of more expensive antimicrobials, greater morbidity, increased costs and excess mortality? ... Do we have to wait until consumers or their lawyers demand implementation of (infection control) programs?" Losing tens of thousands of patients to hospital-made infections now seems to be as acceptable as surrendering entire forests to Asian long-horned beetles or abandoning the Great Lakes to a horde of alien marine invaders.

In the meantime, hospitals are doing a roaring business in germ traffic. One of the latest headline makers is *Acinetobacter baumannii,* a common soil dweller. Twenty years ago it accounted for barely 1.5 percent of all hospital pneumonias. Today it boasts a 7 percent share of hospital-acquired lung destroyers in North America and can be found in 25 percent of all Spanish hospitals. This drug-resistant bug has also become the bane of maimed American soldiers returning from Iraq and is spreading rapidly through U.S. hospitals and their paraplegic centers. About 40 percent of the infected patients die. "It's a big problem," recently admitted Colonel Bruno Petrucelli in the British science journal *Nature:* "Hospitals are dangerous places to be."

Nor are they getting safer. Last year the *Journal of Infection Control and Hospital Epidemiology* published a string of articles with disconcerting titles including "Why Is It That Internists Do Not Follow Guidelines for Preventing Intravascular Catheter Infections?" and "Double, Double, Toil and Trouble: Infections Still Spreading in Long-Term-Care Facilities." Another eyepopper posed a rhetorical question: "Are Antibiotic Resistant Nosocomial Infections Spiraling Out of Control?" The answer was an unqualified yes. Seventy percent of all hospital-acquired infections are now resistant to the standard drugs used to treat them. In addition, most cases of MRSA still go unrecognized and are never isolated in North American hospitals.

The endless proliferation of hospital invaders and their eagerness to expand into the greater community suggest the world's next pandemic might well begin, like SARS, in a crowded emergency department or nursing home. More than a century ago, Florence Nightingale recognized that hospitals were a menace to the public health and called for all kinds of reforms. She also knew that the institution had its limits and wisely concluded that hospitals represented "only an intermediate stage of civilization."

Other medical professionals have reached more radical conclusions about the globe's growing biological instability. They regard the invasion of hospitals as a crisis for human medicine, just as the invasion of livestock has become a crisis for veterinary medicine.

And so they tell black jokes at infectious disease conferences about how modern medicine has breached the boundaries of the human condition and found its nemesis in biological invaders: "The 19th century was followed by the 20th century, which was followed by the ... 19th century."

Epilogue:
The Next Pandemic

"No longer were there individual destinies; only a collective destiny, made of plague and emotions shared by all."
—Albert Camus

S ix years ago the U.S. government staged a mock biological attack in Denver. The simulated exercise, which involved top officials of the state and federal government, cost $3 million. Like two subsequent exercises, it was designed to test the efficacy of emergency and medical plans. The mock attack began on May 17 with the simulated release of an aerosol spray containing bubonic plague. On Day 1, 783 citizens fell ill; on Day 2, hospitals ran out of drugs, and infected travelers popped up in England and Japan; by Day 3, city hospitals were forced to turn away the plague-ridden; by Day 4, there were more than 3,700 cases of infection and 2,000 deaths. "Disease had already spread to other states and countries," recorded one public health expert and observer, Thomas Inglesby. "Competition between cities for the National Pharmaceutical Stockpile had already broken out. It had all the characteristics of an epidemic out of control."

If the next invader is a kin of the bird flu or H5N1, it will go global in weeks and soon be out of control. As it explodes around the world, the invader will find three conditions unique to our era: a just-in-time global economy, unprecedented urban crowding, and unparalleled human mobility that our ancestors might have regarded as miraculous or even divine. These unwitting preparations will shape the next pandemic and drive it into unknown territory.

Ancient societies understood the importance of surge capacity and often devised seven-year plans to withstand famine, drought, and plague. Technological society has mocked such forethought. The essayist Ian Welsh thinks our new economic habits of living like hedonistic grasshoppers as opposed to Aesop's prudent ants will disastrously magnify the pandemic's impact. "Our society, as a whole, has no surge protection, no ability to take shocks. We have no excess beds, no excess equipment, no excess ability to produce vaccines or medicines, nothing. Everybody has worshiped at the altar of efficiency for so long that they don't understand that if you don't have extra capacity you have no ability to deal with unexpected events." The pandemic, then, will test the sustainability of the just-in-time supply chains pioneered by Wal-Mart and other multinationals.

Urban crowding will also shape the course of the invasion. Two hundred years ago, most humans lived in rural areas, and only one city in the world boasted a population of 1 million people: London. But global trade and migration in the last 100 years has changed that picture by concentrating more than half of the world's people in cities. Today 35 metropolises—human feedlots—claim more than 5 million residents each, and hundreds harbor more than 1 million souls each. Tokyo proudly supports 35 million people, Mexico City claims 19 million "official" residents, and London boasts 7 million inhabitants. The most crowded slums of Mumbai, Delhi, or Nairobi nurture population densities exceeding 80,000 residents per square kilometer (200,000 residents per square mile).

This kind of concentration is dry tinder to viral sparks, and it guarantees the "superdiffusion" of any viral invader. "We are all, so to speak, sitting in the waiting room of an enormous clinic, elbow to elbow with the sick of the world," writes the historian Alfred Crosby, Jr. A 2004 report by the CIA's National Intelligence Council predicts that a "pandemic in the megacities of the developing world with poor health-care systems" could spread with devastating effect and "derail globalization."

Historically pandemics have never been kind to urban dwellers. According to the Greek historian Thucydides, the world's first recorded pandemic (known as the plague of Athens) killed almost a third of the population with what was likely typhoid. It arrived with merchants from Ethiopia. The Antonine plague (the plague of Galen), a visitation of measles or smallpox carried by returning imperial troops, emptied the city of Rome in AD 165 and dispatched two emperors. The plague of Justinian, an invasion of bubonic plague spread by African grain ships, silenced 40 percent of the population of Constantinople in AD 541. In the 14th century, the Black Death often killed off half the residents of urban slums. The 1918–19 influenza pandemic buried more than 50 million people. Most were either urban dwellers or soldiers packed into camps that resembled slums.

Unlike these past pandemics, which rarely traveled faster than 2 kilometers (1.2 miles) a day, the next one will break all speed records. Long before we suspect its presence, the invader will be traveling thousands of kilometers or miles a day by planes, buses, or trains. It will then spread out over shorter distances by cars, bicycles, subways, inline skates, and skateboards. Soccer moms may innocently drive the pandemic across every major North American city.

The dying will begin with a careless sneeze that will propel as many as 100,000 viral particles from one end of an airplane cabin to the other at 128 kilometers (80 miles) an hour. The virus will enter the throat and upper lungs of a mobile trader or a tanned tourist and then hijack the cells to make more viral invaders. Two to seven days later, the infected will feel fluish. If each person infects just two others, then one infection will become 1,024 infections within 30 days, which means the pandemic won't become a serious conflagration. But if it finds superspreaders and hits a reproductive rate of greater than three, the invader will change history. The first North American cases will probably appear in either Vancouver or Los Angeles, key centres in the trade with Asia.

Initial symptoms may mimic the ordinary aches and pains of the flu or offer something altogether different. During the 1918–19 pandemic, influenza triggered a storm in the immune system of young adults by flooding their lungs with anti-inflammatory chemicals and suffocating their airways. This exuberant immune response (a "cytokine storm") had many disturbing manifestations. Some people watched their fingers and genitals turn black, while others smelled their bodies decaying before death declared itself. Some went to bed with a pounding headache and never woke up. Others survived but remained brain damaged for life.

Medical experts generally agree that if the pandemic is mild, the flu will infect between 25 and 30 percent of the world's population and kill 2 million to 7 million people. Such an event will not be a demographic disaster or an economic catastrophe. But a nasty invader along the lines of the 1918–19 strain would leave a bitter wake. Any influenza strain that targets 20- to 40-year-olds and tricks the immune system into searing the lungs could yield global death tolls as high as 300 million. The death rate among pregnant women, the most vulnerable, could approach a horrific rate of 70 percent. Such a rare and unique event would change the world.

According to reports by the U.S. Congressional Budget Office and BMO Nesbitt Burns, a severe pandemic with a 2.5 percent death rate will shock the economy and turn the health-care system upside down. It will infect more than 100 million people in North America and kill at least 2 million. Such an event will rapidly erase North America's mirage of security. With more than a third of the workforce sick, or caring for the sick, vital public services will suffer or collapse. Water treatment technicians will be too sick to purify urban water supplies, while garbage collectors will be too listless to pick up the trash. A short-age of electrical workers could trigger a series of power outages in a decrepit system already prone to blackouts. Other invaders, from rats to flies, will seize the day to feed on growing mountains of human waste—and the dead.

Debates about the vaccines and drugs will obsess the media, but to no avail. Scientists agree that it will probably take four to six months to produce an effective antiviral vaccine. Limited supplies mean that members of the military, the government, and the health-care system will get the shots first. Given the real possibility of adverse reactions caused by vaccine contamination, a public health controversy like the one that blackened the effort to protect against anthrax might well engulf the flu vaccine campaign. Although drugs such as Tamiflu (a pill) and Relenza (an inhalation mist) may be available in limited quantities, their profligate use will quickly generate drug-resistant strains. Both drugs are totally ineffective if not taken within 48 hours of the first symptoms. Economists calculate that it would take $14 billion and another 10 years to produce enough antiviral drugs for 20 percent of the world's population.

As a result, governments will champion low-tech messages such as "Coughs and sneezes spread diseases." They will tell people to wash their hands, cover their mouths, and avoid crowds. They will command the sick to remain in bed and advise people not to touch their eyes, nose, and mouth. Basic hygiene and basic manners will become matters of life and death.

In response to a severe pandemic, governments will close borders and reluctantly impose quarantines on the sick. If voluntary quarantines fail, authorities will close schools and universities and turn them into temporary morgues or hospitals. They will forbid public gatherings and sporting events and shut down movie complexes. Public transit will come to a standstill. People will fret in their homes and pray for the best. Some will die flicking channels or listening to music on their iPods.

Within a week of the invasion, people will have trouble buying food and medical supplies. The cemeteries will overfill and local meatpackers will store the dead in refrigerated trucks. Rumors that cats and dogs may spread the invader will prompt urban animal massacres. (Cats, pigs, or cattle may well become influenza reservoirs.) Entire police precincts, fearful of hauling away the dead and weary of nailing red influenza signs on doors at the homes of the infected, will report sick and not answer calls. Ambulance drivers will cough and ache with fever and park their vehicles. Ailing truck drivers will abandon their rigs and perish in coffee shops away from home. The media will advertise critical shortages of doctors, nurses, gravediggers, and truck drivers.

Grocery stores will ration food and station armed guards at the doors during a serious outbreak. Doctors will charge exorbitant fees for home visits. A black market in alternative therapies, illegal vaccines, and Chinese herbal remedies will flourish. Few Boy Scouts will deliver groceries to the infirm and the housebound because, in contrast to 1918–19, the Boy Scouts are now a spent organization.

Hospitals will quickly see more sick people than they can treat. The United States has beds for 1 million people, but even a minor pandemic would create at least 5 million infections needing hospitalization. So hallways will fill up with gurneys overflowing with the unwell or the "worried well." Within a week, most hospitals and clinics will run out of beds, gloves, face masks, disinfectants, antiviral drugs, and antibiotics. The demand for mechanical ventilators for those with failing lungs will quickly exceed the supply. Most people needing a ventilator to breathe will not find one and will die.

A severe pandemic, then, will quickly disarm and neutralize hospitals. Health workers and "first responders," either dismayed by the chaos or frightened by the lethality of the invader, will refuse to show up at work. When bubonic plague appeared in the industrial port of Surat, India, in 1994, 70 percent of the private doctors fled the city, saying "nothing could be done." In England nearly 40 percent of the nursing staff will go missing, under some scenarios. And then the second and third waves will hit.

As local governments try to implement their pandemic preparedness plans, they will rudely discover what Thomas Glass, an American public health expert, warned years ago: "Disaster planning does not go as planned." Communities under biological attack often respond collectively and creatively, argues Glass. They normally do so without emergency professionals in the neighborhood. In the absence of white coats, they will form spontaneous groups that will establish their own roles, rules, and leaders. Government plans that don't recognize and support the public as first responders and as the only auxiliaries that truly matter will fail and fail miserably. During the 1918–19 pandemic, the public fed and nursed the sick at home and then buried the dead. The next severe pandemic will be no different.

Each and every city will respond to the plague in its own way. Singapore will maintain its martial discipline, but Moscow will fall into chaos. New York, seasoned by 9/11, will rise above the fray to be New York. Cities that have invested in their citizenry and communities will pass through the days of the dead with grace. Others will fall apart like New Orleans. Wherever the government has ignored its poor and undermined its communities, conflict and chaos will rule the streets. There will be many New Orleanses.

During the 1918–19 pandemic, a combination of leadership, citizenship, and neighborliness sharply set apart those cities that overcame the invasion from those that succumbed to barbarism. Philadelphia, a city led by a crooked mayor and inept public health authorities, suffered horribly. It abandoned its teeming black neighborhoods and dumped the dead in piles outside police stations. While hundreds died every day, the authorities issued empty press releases that read: "There is no cause for panic or alarm."

Meanwhile San Francisco, which had recovered from a devastating earthquake with community-based organizations, proudly marched on in spite of all the dying. After the schools closed, city teachers volunteered to serve as nurses, gravediggers, and telephone operators. Unlike

Philadelphia and its outrageous spin doctors, Frisco authorities focused on containing the invader as opposed to corralling panic. Every citizen wore a mask and pitched in.

"Most communities were woefully unprepared for the health crisis they faced," writes the American historian Jim Higgins. "Those cities that passed muster had been building a strong medical infrastructure for decades and had sound public health policies based more upon science than politics. I'm not sure that's the case today." In fact, most local and national governments in North America will soon discover the folly of underfunding and neglecting their public health departments.

A severe pandemic will shock the global economy with the same force as a major economic depression. Free-market ideologues have never prepared entrepreneurs for financial busts, let alone biological ones. In 2005 the Asian Development Bank predicted that a pandemic might cost the region anywhere between $100 billion and $300 billion in lost trade and investment. The world's third-largest bank, HSBC, expects that half of its workforce of 253,000 will fall ill, but it plans to carry on from employees' homes via video links and teleconference facilities. No one really expects banks to keep their cash machines restocked on a regular basis.

Sherry Cooper, a Canadian economist at BMO Nesbitt Burns, predicts that a severe pandemic will essentially close restaurants, casinos, sports arenas, hotels, airlines, cruise ships, tourist resorts, and dental offices for months on end. It will also cook the global poultry trade and deep-fry many fast-food outlets such as Kentucky Fried Chicken. She thinks the world's 70,000 multinationals should expect a third of their employees to be sick or absent at any one time. Given that no nation is self-sufficient in basic goods anymore, she foresees major complications involving either crippling shortages or huge wastage of untrucked goods at closed borders.

At some point the pandemic will meet and embrace other invaders. The sheer number of immunocompromised Africans with HIV might accelerate the evolution of influenza and promote deadly mutations. North America's aging baby boom generation (some 78 million people) might add fuel to the fire as well. (By 2050 nearly a quarter of the world's people will be over 60 years of age. That represents a new monoculture ripe for microbial invasions.) Every pandemic plays with the material at hand.

If the death rate eventually exceeds 2.5 percent of the population (an unlikely event), prepare for the revival of old plague moralities. Thucydides saw it among ordinary Athenians. And Boccaccio saw it among rich Florentines of the 14th century who fled their plague-ridden city to country villas, where they thought a lot about food and sex: "A dead man was then of no more account than a dead goat would be today."

When death becomes a constant companion, people simply abandon their old virtues and inhibitions. "So they resolved to spend quickly," wrote Thucydides, "and enjoy themselves regarding their lives and riches as alike things of a day. Perseverance in what men called honor was popular with none. It was so uncertain whether they would be spared to attain the object; but it was settled that present enjoyment and all that contributed to it, was both honorable and useful."

In the end, it will all come down to family and community, the only first responders that have ever mattered in history. Families that love one another and have prepared for biological chance by stockpiling at least a six-week supply of water, dry goods, soap, and medical supplies will fare better in the darkest days than those who surrender their fate to government preparedness plans. Rural communities that still value self-reliance and neighborliness will pass through the ordeal with greater dignity than urban monocultures of wealth or poverty. Ian Welsh has emphasized that "one of the most important predictors for survival will be the strength of social ties to others." Families, too, will relearn the importance of disciplined care and prudent habits. During a pandemic, "the man who infects hardly anyone is the man who has the fewest lapses of attention," says Camus.

After the Great Mortality, other invaders will take advantage of influenza's toxic sequelae: broken public health systems, damaged lungs, and wounded governments. Drug-resistant tuberculosis will exploit the world's weakened lungs and ailing immune systems to gain new ground. Those hospitalized for influenza treatment will inadvertently spread MRSA and *C. difficile* and other by-products of industrial medicine to their communities. HIV, which has morphed from 2 to 400 strains in the last 20 years, will surge forward in the wake of public health programs decimated by influenza. Unattended livestock and contaminated water systems will attract their own microbial looters. Biological aftershocks will shake economies and societies for years to come. Climate change will be the joker in this black deck of biological mayhem.

When the dead have been burned in industrial-sized pyres and the stores have tentatively reopened, some of us will belatedly remember that only human love can overpower fear. Others will praise the spirit of human resilience. Most will thank God for sparing their families. Fossil fuel prices will fall for several years, and a record number of people will apply for new life insurance policies. The Jazz Age followed the 1918–19 pandemic, and the Renaissance bloomed after the Black Death. We may find new balance and emerge healthier and wiser or collapse into a new Dark Age.

Ultimately, a severe pandemic might encourage us to rethink the deadly pace of globalization and biological traffic in all living things. Ian Welsh believes that people will travel less, trade less, and care more about the public health in a post-pandemic world. "There will be a lot less foreign goods on the store shelves," he suggests, and he is probably right. After the Great Mortality we might question the crowding of livestock into factory farms, the separation of veterinary and human medicine, the pollution of our waters, our dramatic fouling of the weather, the vulnerability of monocultures, and the globalization of everything. Maybe we will learn that we can't liberalize trade without liberating biology in unpredictable ways.

Long after the monotony of deprivation and separation, the survivors of the Great Mortality will kiss their loved ones each night and hold them tight. Then they will light candles, true plague light, and pray for deliverance from more invaders. The humbled will be thankful, as Albert Camus once was, for what pandemics have always taught those receptive to biological instruction: "There are more things to admire in men than to despise."

A Canticle for Local Living

M any of the biological invasions documented in this book could have been prevented. In fact, it's much easier and cheaper to identify a biological hitchhiker and close down its pathway than it is to stop a runaway pandemic. But it requires responsible trade with responsible controls, something that free-trading governments are reluctant to do.

Yet prosperous Australia and New Zealand have decided that liberal trade requires some respectful restraints. After being overwhelmed by biological invasions in the 19th century, both nations have kept modern-day invaders in check with well-funded programs that license, inspect, and often quarantine certain foreign goods. In short, these countries carefully identify troublesome hitchhikers and then quickly intervene to shut down the pathways. An increasing number of economists now argue that nations with lots of people, trade, and livestock must work harder at identifying potential global trade pathways (from ballast water to animal feed) in order to avoid ruinous plagues. Sensible trade restrictions on agricultural and timber goods could save a forest, a herd of cattle, or even an entire city. The exotic pet industry is an outright public health hazard and should probably be banned.

But waiting for governments to do the right thing can be a hazardous enterprise and a test of patience. Citizens can fill that void by simply changing the way they eat, buy, and live. If unrestrained global trade in all living things has created unparalleled biological mayhem, then maybe it's time to act and think more locally. Maybe it's time to abandon what Wendell Berry calls our "arrogant ignorance" about the impacts of our global purchases and return to personal and local virtues that question bigness and power. Maybe it's time to learn a new canticle for creation that encourages, as Saint Francis did, humility.

1. "First, say to yourself what you would be; then do what you have to do."—Epictetus

2. Ask the questions that Charles Elton and Wendell Berry ask: What is here? What will nature permit us to do here? What will nature help us to do here?

3. Study the state motto of Hawaii: "The life of the land is perpetuated in righteousness."

4. Grow an interest in agriculture, farmer's markets, slow food, and organic production.

5. Eat only vegetables and fruits grown locally or organically.

6. Eat wild fish sparingly and only that caught locally.

7. Eat beef that is grass fed.

8. Laugh loudly at the dinner table.

9. Advocate the proper separation of trade promoters and food protectors in agricultural departments.

10. Praise, as Saint Francis did, Sister Water, "so useful, humble, precious, and pure."

11. Keep feedlots, cities, and industry out of our watersheds.

12. Stick to known paths and trails.

13. Avoid hospitals. Clean your shoes and wash your clothes after every visit.

14. Always wash your hands.

15. Advocate for a 19th-century ideal: health care in the home with a minimum of professionals by your side.

16. Shun antibiotics.

17. Regard exotic pets such as turtles, snakes, and monkeys as bad spouses: Sooner or later they'll bite you. Don't buy them, don't trade them, and don't release them into the wild. Return them to the pet shop.

18. Plant local herbs and flowers. Demand that nurseries label noxious invasive exotics. Without controls, these invaders can transform your neighborhood with the same clout as urban gangs.

19. Travel lightly and infrequently.

20. Spend time outdoors, for what you see will soon be changed by invasions.

21. Buy only locally produced wood products and no wooden items from Asia or Africa.

22. Work to prohibit the import of entire plants or living plant parts. Tolerate the imports of only small amounts of seed or clones grown in vitro, free of bacteria and fungi.

23. Advocate the treatment of all imported logs, lumber, and chips with high heat to eliminate all invasive species. Demand that unprocessed wood be phased out as a packing material.

24. Wash your car before traveling out of state, province, or country.

25. Accept responsibility for the biological consequences of one's actions. The Republican ecologist Garrett Hardin put it best: "Self interest urges individuals to evade responsibility whenever they possibly can. The more distant the monitor, the more feasible evasion becomes. Globalization favors evasion. The wise rule to follow should be plain: Never globalize a problem if it can possibly be dealt with locally."

26. True homeland security does not come from guns, border guards, or ID papers. Sound national defense and biological security, says Wendell Berry, can come only from a certain rootedness: "widespread, settled, thriving local communities, each having a proper degree of independence, living so far as possible from local sources, and using its local sources with stewardly care."

27. Be prepared for surprises and abrupt die-offs. When the whole world becomes a port of call, death becomes an obligatory caller.

28. Question the promoters of globalization; they don't hold degrees in biology or ecology. Ask them where they plan to retire.

29. Pick weeds whenever you can. It's good for the soul.

30. Conserve fossil fuels. If we all burn fewer hydrocarbons, we might be able to keep the global thermostat at a temperature that will conserve the world's biological diversity and ultimately our greater health.

31. Buy less and live more.

32. "Learn to live happily with less than we can dream of."
    —Garrett Hardin

Appendix:
14 Steps You Can Take to Reduce Your Risk of a Hospital Infection*

1. Ask that hospital staff clean their hands before treating you. This is the single most important way to protect yourself in the hospital. If you're worried about being too aggressive, just remember your life could be at stake. All caregivers should clean their hands before treating you. Alcohol-based hand cleaners are more effective at removing most bacteria than soap and water.[1] Do not hesitate to say the following to your doctor or caregiver: "Excuse me, but there's an alcohol dispenser right there. Would you mind using that before you touch me, so I can see it?" Don't be falsely assured by gloves. Gloves more often protect staff than patients. If caregivers have pulled on gloves without cleaning their hands first, the gloves are already contaminated before they touch you.[2]

2. Before your doctor uses a stethoscope to listen to your chest, ask that the diaphragm (or flat surface of the stethoscope) be wiped with alcohol. Numerous studies show that stethoscopes are often contaminated with *Staphylococcus aureus* and other dangerous bacteria, because caregivers seldom take the time to clean them in between patient use.[3] The American Medical Association recommends that stethoscopes routinely be cleaned for each patient. The same precautions should be taken for many other commonly used pieces of equipment too.

3. Ask visitors to clean their hands and avoid sitting on your bed.[4]

4. If you need surgery, choose a surgeon with a low infection rate. Surgeons know their rate of infection for various procedures. Ask for it. If they won't tell you, consider choosing another surgeon. You should be able to compare hospital infection rates too, but that information is almost impossible to get. That is why RID is working hard for hospital infection report cards in every state.

5. Beginning one week before surgery, shower frequently with Chlorhexidine soap. Various brands can be found at drug stores. This will help remove any dangerous bacteria you may be carrying on your own skin.

6. Ask your surgeon to have you tested for *Staphylococcus aureus* at least one week before you come into the hospital. The test is simple, usually just a nasal swab. About one third of people carry *Staphylococcus aureus* on their skin, and if you are one of them, extra precautions can be taken to protect you from infection, to give you the correct antibiotic during surgery, and to prevent you from transmitting bacteria to others.

7. On the day of your operation, remind your doctor that you may need an antibiotic one hour before the first incision. For many types of surgery, a pre-surgical antibiotic is the standard of care, but it is often overlooked by busy hospital staff.[5]

8. Ask your doctor about keeping you warm during surgery. Operating rooms are often kept cold for the comfort of the staff, but research shows that for many types of surgery, patients who are kept warm resist infection better.[6] There are many ways to keep patients warm, including special blankets, hats and booties, and warmed IV liquids.

9. Do not shave the surgical site. Razors can create small nicks in the skin, through which bacteria can enter. If hair must be removed before surgery, ask that clippers be used instead of a razor.[7]

10. Ask that your surgeon limit the number of personnel (including medical students) in the operating room. Every increase in the number of people adds to your risk of infection.[8]

11. Ask your doctor about monitoring your glucose (sugar) levels continuously during and after surgery, especially if you are having cardiac surgery. The stress of surgery often makes glucose levels spike erratically. New research shows that when blood glucose levels are tightly controlled to stay between 80–110 mg/unit, heart patients resist infection better. Continue monitoring even when you are discharged from the hospital, because you are not fully healed yet.[9]

12. Avoid a urinary tract catheter if possible. It is a common cause of infection. The tube allows urine to flow from your bladder out of your body. Sometimes catheters are used when busy hospital staff don't have time to walk patients to the bathroom. Ask for a diaper or bed pan instead. They're safer.[10]

13. If you must have an IV, make sure that it is inserted and removed under clean conditions and changed every 3 to 4 days. Intravenous catheters, or IVs, are a common source of infection and are not always necessary. If you need one, insist that it be inserted and removed under clean conditions, which means that your skin is cleaned at the site of insertion, and the person treating you is wearing clean gloves. Alert hospital staff immediately if any redness appears.

14. If you are planning to have your baby by Cesarean section, take the steps listed [earlier] as if you were having any other type of surgery.

NOTES:

1. Vincent JL, Bihari DJ, Suter PM, et al. The prevalence of nosocomial 380 INFECTION CONTROL AND HOSPITAL EPIDEMIOLOGY May 2003 infection in intensive care units in Europe: results of the European Prevalence of Infection in Intensive Care (EPIC) Study. EPIC International Advisory Committee. *JAMA* 1995; 274: 639–644.

2. Cars O, Molstad S, Melander A. Variation in antibiotic use in the European Union. *Lancet* 2001; 357: 1851–1853.

3. Frank MO, Batteiger BE, Sorensen SJ, et al. Decrease in expenditures and selected nosocomial infections following implementation of an antimicrobial-prescribing improvement program. *Clinical Performance and Quality Healthcare* 1997; 5: 180–188.

4. Fukatsu K, Saito HK, Matsuda T, Ikeda S, Furukawa S, Muto T. Influences of type and duration of antimicrobial prophylaxis on an outbreak of methicillin-resistant *Staphylococcus aureus* and on the incidence of wound infection. *Arch Surg* 1997; 132: 1320–1325.

5. Batteiger BE. Personal communication. Indianapolis: Indiana University; 2001.

6. Back NA, Linnemann CC Jr, Staneck JL, Kotagal UR. Control of methicillin-resistant *Staphylococcus aureus* in a neonatal intensive-care unit: use of intensive microbiologic surveillance and mupirocin. *Infect Control Hosp Epidemiol* 1996; 17: 227–231.

7. Barrett FF, McGehee RF, Finland M. Methicillin-resistant *Staphylococcus aureus* at Boston City Hospital. *N Engl J Med* 1968; 279: 441–448.

8. Boyce JM. Are the epidemiology and microbiology of methicillin-resistant *Staphylococcus aureus* changing? JAMA 1998; 279: 623–624.

9. Brumfitt W, Hamilton-Miller J. Methicillin-resistant *Staphylococcus aureus*. *N Engl J Med* 1989; 320: 1188–1196.

10. Naimi TS, LeDell KH, Boxrud D, et al. Epidemiology and clonality of community acquired methicillin-resistant *Staphylococcus aureus* in Minnesota, 1996-1998. *Clin Infect Dis* 2001; 33: 990–996.

Selected Bibliography

This is only a partial bibliography, listing about 20 key references for each chapter. *Pandemonium* took three years to write and is based on more than 3,000 books and articles.

CHAPTER ONE: AVIAN FLU: BARBARIANS IN THE COOP

Crosby, Alfred, Jr. *America's Forgotten Pandemic: The Influenza of 1918.* Cambridge: Cambridge University Press, 2003.

Davis, Karen. *The Avian Flu Crisis in Canada: Ethics of Farmed-Animal Disease Control.* United Poultry Concerns, February 25, 2005. www.upc-online.org/slaughter/22805karenflu.htm.

Davis, Mike. *The Monster at Our Door: The Global Threat of Avian Flu.* New York: New Press, 2005.

Dekich, Mark. "Broiler Industry Strategies for Control of Respiratory and Enteric Diseases." *Poultry Science* 77 (1998): 1176-80.

Food and Agriculture Organization (FAO). *FAO Workshop on Social and Economic Impacts of Avian Influenza Control, 8-9 December 2004.* http://www.fao.org/ag/againfo/subjects/documents/AIReport.pdf.

Food and Agriculture Organization (FAO). "High Geographic Concentration of Animals May Have Favoured the Spread of Avian Flu." News release, January 28, 2004. http://www.fao.org/newsroom/en/news/2004/36147/index.html.

Food and Agriculture Organization (FAO) and World Organisation for Animal Health (OIE). *A Global Strategy for the Progressive Control of Highly Pathogenic*

Avian Influenza (HPAI). Rome and Paris: FAO and OIE, November 2005. http://www.fao.org/ag/againfo/subjects/documents/ai/HPAIGlobalStrategy 31Oct05.pdf.

Goudsmit, Jaap. *Viral Fitness: The Next SARS and West Nile in the Making*. New York: Oxford University Press, 2004.

Halvorson, David. "Modern Disease Control for a Modern Poultry Industry." University of Minnesota College of Veterinary Medicine, [2004]. www.cvm.umn.edu/img/assets/19565/AI_control.pdf.

Lee, Chang-Won, et al. "Effect of Vaccine Use in the Evolution of Mexican Lineage H5N2 Avian Influenza Virus." *Journal of Virology* 78, no. 15 (August 2004): 8372-81.

Mackenzie, Debora. "Genes of Deadly Bird Flu Reveal Chinese Origin." *New Scientist*, February 6, 2006.

McNeil, Donald, Jr. "A New Deadly, Contagious Dog Flu Virus Is Detected in Seven States." *New York Times*, September 22, 2005.

Morris, R.S., and R. Jackson. *Epidemiology of H5N1 Avian Influenza in Asia and Implications for Regional Control* [Massey Report]. Rome: FAO, April 2005.

Nicholson, K.B., et al. *Textbook of Influenza*. Oxford: Blackwell Science, 1998.

Nierenberg, Danielle, and Leah Garcés. *Industrial Animal Agriculture—The Next Global Health Crisis?* London: World Society for the Protection of Animals, 2004.

Royal Society for the Protection of Animals. *Paying the Price: The Facts About Chicken Reared for Their Meat*. London: RSPCA, 2005. http://www.rspca.org.uk.

Slingenbergh, J., et al. "Ecological Sources of Zoonotic Diseases." *OIE Scientific and Technical Review* 23, no. 2 (2004): 467-84.

Specter, Michael. "Nature's Bioterrorist." *New Yorker*, February 28, 2005.

Webster, R.G., and D.J. Hulse. "Microbial Adaptation and Change: Avian Influenza." *OIE Scientific and Technical Review* 23, no. 2 (2004): 453-65.

For excellent sources on the progress and evolution of H5N1, see Henry Niman's Web site, http://www.recombinomics.com/whats_new.html, or the Centre for Infectious Disease Research and Policy, http://www.cidrap.umn.edu/. See also the Flu in China Web site, http://www.flu.org.cn. Crawford Kilian's useful blog is at http://crofsblogs.typepad.com/h5n1/.

CHAPTER TWO: JUGGLING SPECIES

Bright, Chris. *Life Out of Bounds.* New York: W.W. Norton, 1998.

Campbell, Faith, et al. *Fading Forests 2.* Washington, DC: American Lands Alliance, 2002.

Claudi, Renata, et al. *Alien Invaders in Canada's Waters, Wetlands and Forests.* Ottawa: Natural Resources Canada, 2002.

Coatsworth, John. "Globalization, Growth, and Welfare in History." In *Globalization: Culture and Education in the New Millenium,* edited by Marcelo Suarez Orozco and Desiré Baolian Qin-Hilliard, 38-55. Berkeley: University of California Press, 2004.

Crosby, Alfred, Jr. *Germs, Seeds and Animals: Studies in Ecological History.* New York: M.E. Sharpe, 1994.

Cunningham, A., et al. "Pathogen Pollution: Defining a Parasitological Threat to Biodiversity Conservation." *Journal of Parasitology* 89 (2003): s78-s83.

Daszak, Peter, et al. "Emerging Infectious Diseases of Wildlife—Threats to Biodiversity and Human Health." *Science* 287, January 21, 2000.

Elton, Charles. *The Ecology of Invasions by Animals and Plants.* Chicago: University of Chicago Press, 1958.

Hawthorne, Michael. "Exotic Invaders Threaten Environment, Economy." *Columbus Dispatch,* October 26, 2003.

Holeck, Kristen, et al. "Bridging Troubled Waters: Biological Invasions, Transoceanic Shipping and the Laurentian Great Lakes." *BioScience* 54, no. 10 (October 2004): 919-29.

Karesh, William, et al. "Wildlife Trade and Global Disease Emergence." *Emerging Infectious Diseases* 11, no. 7 (July 2005).

Lederberg, Joshua. "Infectious Disease as an Evolutionary Paradigm." *Emerging Infectious Diseases* 3, no. 4 (December 1997).

Loope, L.L., and F.G. Howarth. "Globalization and Pest Invasion: Where Will We Be in Five Years?" In *Proceedings of the International Symposium on Biological Control of Arthropods,* Honolulu, Hawaii, 14-18 January 2002, edited by R.G. Van Driesche, 34-39. Morgantown, WV: U.S. Department of Agriculture Forest Service. Available at http://www.bugwood.org/arthropod/day1/loope.pdf.

MacIsaac, Hugh, et al. *Economic Impacts of Invasive Nonindigenous Species in Canada: A Case Study Approach.* Ottawa: Auditor General of Canada, May 2002.

McKinney, Michael, and J.L. Lockwood. "Biotic Homogenization: A Few Winners Replacing Many Losers in the Next Mass Extinction." *Tree,* November 1999.

McMichael, A.J. "Environmental and Social Influences on Emerging Infectious Diseases: Past, Present and Future." *Philosophical Transactions of the Royal Society of London* B 359 (2004): 1049-58.

McNeely, Jeffrey. *The Great Reshuffling: Human Dimensions of Invasive Alien Species.* Cambridge, UK: World Conservation Union (IUCN), 2001.

Mills, Edward, et al. "Exotic Species and the Integrity of the Great Lakes: Lessons from the Past." BioScience 44, no. 10 (November 1994): 666-76.

Murphy, Frederick. "The Threat Posed by the Global Emergence of Livestock, Food-borne and Zoonotic Pathogens." *Annals of the New York Academy of Sciences* 894 (1999): 20-27.

Normile, Dennis. "Expanding Trade with China Creates Ecological Backlash." *Science* 306, November 5, 2004.

Simberloff, Daniel. "Impacts of Introduced Species in the United States." *Consequences* 2, no. 2 (1996).

———. "The Politics of Assessing Risk for Biological Invasions: The USA as a Case Study." *Trends in Ecology and Evolution* 20, no. 5 (May 2005).

CHAPTER THREE: LIVESTOCK PLAGUES

Campbell, David, and Robert Lee. "Carnage by Computer: The Blackboard Economics of the 2001 Foot and Mouth Epidemic." *Social & Legal Studies* 12, no. 4 (2003): 425-59. http://sls.sagepub.com/cgi/content/abstract/12/4/425.

Cooke, B.D. "Rabbit Haemorrhagic Disease: Field Epidemiology and the Management of Wild Rabbit Populations." *OIE Scientific and Technical Review* 21, no. 2 (2002): 347-58.

Doward, Jamie. "The Illegal Meat Trade That Can Bring a Deadly Virus to Our High Streets." *Observer,* May 8, 2005.

Fleming, George. *Animal Plagues: Their History, Nature and Prevention.* London: Chapman and Hall, 1871.

Kerbis-Peterhans, J.C., and T.P. Gnoske. "The Science of 'Man-eating' among Lions Panthera leo with a Reconstruction of the Natural History of the 'Man-eaters of Tsavo.'" *Journal of East Africa Natural History* 90, nos. 1-2 (2001): 1-40.

Kjekshus, Helge. *Ecology Control and Economic Development in East African History: The Case of Tanganyika, 1850-1950.* London: Heinemann, 1977.

Kouba, Vaclav. "Globalization of Communicable Animal Diseases—A Crisis of Veterinary Medicine." *Acta Veterinaria Brno* 72 (2003): 453-60.

———. "Public Service Veterinarians Worldwide: A Quantitative Analysis." *Acta Veterinaria Brno* 74 (2005): 455-61.

Machado, Manuel. *Aftosa: A Historical Survey of Foot-and-Mouth Disease and Inter-American Relations.* Albany: State University of New York Press, 1969.

Miller, Jonathan. "A Peasant Revolts." *Sunday Times* (London), April 29, 2001.

Mort, Maggie, et al. "Psychosocial Effects of the 2001 UK Foot and Mouth Disease Epidemic in a Rural Population: Qualitative Diary Based Study." *British Medical Journal* 331 (November 26, 2005): 1234-39.

Nelson, Robert. "Environmental Colonialism: 'Saving' Africa from Africans." Paper presented at the World Summit on Sustainable Development, Johannesburg, South Africa, August 25, 2002.

Phoofolo, Pule. "Epidemics and Revolutions: The Rinderpest Epidemic in Late Nineteenth Century Southern Africa." *Past and Present,* no. 138 (February 1993): 112-43.

Saunders, L. "Virchow's Contributions to Veterinary Medicine: Celebrated Then, Forgotten Now." *Veterinary Pathology* 37 (2000): 199-207.

Scott, Gordon. "The Murrain Now Known as Rinderpest." *Newsletter of the Tropical Agriculture Association* (U.K.) 20, no. 4 (2000): 14-16.

Stockdale, Christopher. "An Effective National Animal Disease Strategy Is Dependent on Frequent Objective Review and the Allocation of Appropriate Resources to Carry Out That Strategy." Dissertation, Royal Agricultural College, Cirencester, 2003.

Walters, Mark. *Six Modern Plagues and How We Are Causing Them.* Washington, DC: Island Press, 2003.

Whitlock, Ralph. *The Great Cattle Plague: An Account of the Foot-and-Mouth Epidemic of 1967-8.* London: Barker, 1968.

Woods, Abigail. *A Manufactured Plague: The History of Foot-and-Mouth Disease in Britain.* London: Earthscan: 2004.

———. "Slaughter of the Innocuous." *Times* (London), March 1, 2001.

For extensive archival material on the foot-and-mouth epidemic in Britain, see http://www.warmwell.com/.

For excellent updates on biological spillovers from livestock to wildlife and vice versa, see the Consortium for Conservation Medicine at http://www.conservationmedicine.org/.

The World Organisation for Animal Health (OIE) offers a number of comprehensive documents online at http://www.oie.int/eng/en_index.htm. See also Vaclav Kouba's informative Web site devoted to animal population health: http://www.cbox.cz/vaclavkouba/.

CHAPTER FOUR: THE TRIUMPHANT PRION

Arjona A., et al. "Two Creutzfeldt-Jakob Disease Agents Reproduce Prion Protein-Independent Identities in Cell Cultures." *Proceedings of the National Academy of Sciences* 101 (2004): 8768-73.

Bollinger, Trent, et al. *Chronic Wasting Disease in Canadian Wildlife: An Expert Opinion on the Epidemiology and Risks to Wild Deer.* Saskatoon, SK: Canadian Cooperative Wildlife Health Centre, July 2004.

Brown, Paul, and Raymond Bradley. "1755 and All That: A Historical Primer of Transmissible Spongiform Encephalopathy." *British Medical Journal* 317 (December 1998): 19-26.

European Parliament. Temporary Committee of Inquiry into BSE. *Report.* Manuel Medina Ortega, rapporteur. February 7, 1997. http://www.euro parl.eu.int/conferences/bse/a4002097_en.htm.

Goldenberg, Suzanne. "Culture of Indifference Leaves America Open to BSE." *Guardian,* January 12, 2004.

Kimball, Ann Marie, et al. "Trade Related Infections: Farther, Faster, Quieter." *Globalization and Health* 1, no. 3 (April 22, 2005).

Klitzman, Robert. *The Trembling Mountain: A Personal Account of Kuru, Cannibals and Mad Cow Disease.* London: Perseus Books Group, 1998.

Lacey, Richard. "The Ministry of Agriculture—The Ministry of Truth." *Political Quarterly* 68 (July-September 1997).

Manuelidis, L. "Transmissible Encephalopathies: Speculations and Realities." *Viral Immunology* 16, no. 2 (2003): 123-39.

Matthews, D. et al. "The Potential for Transmissible Spongiform Encephalopathies in Non-ruminant Livestock and Fish." *OIE Scientific and Technical Review* 22, no. 1 (2003): 283-96.

Nikiforuk, Andrew. "Beef with Industry." *Avenue Magazine,* June 2005.

O'Brien, Michael. "Have Lessons Been Learned from the UK Bovine Spongiform Encephalopathy (BSE) Epidemic?" *International Journal of Epidemiology* 29 (2000): 730-33.

Orr, Joan, and Mary Ellen Starodub. "Risk Assessment of Transmissible Spongiform Encephalopathies in Canada." Draft Report for Health Canada, June 30, 2000.

Rampton, Sheldon, and John Stauber. *Mad Cow USA: Could the Nightmare Happen Here?* Monroe, ME: Common Courage Press, 1997.

Rhodes, Richard. *Deadly Feasts: Tracking the Secrets of a Terrifying New Plague.* New York: Simon & Schuster, 1997.

Schwartz, Maxime. *How the Cows Turned Mad: Unlocking the Mysteries of Mad Cow Disease.* Los Angeles: University of California Press, 2001.

Supervie, Virginie, et al. "The Unrecognized French BSE Epidemic." *Veterinary Research* 35 (2004): 349-62.

United Kingdom. BSE Inquiry. *Report.* October 2000. http://www.bse inquiry.gov.uk/.

United States. General Accounting Office. *Mad Cow Disease: Improvement in the Animal Feed Ban and Other Regulatory Areas Would Strengthen US Prevention Efforts.* Report to Congressional Requesters. Washington, DC: GAO, January 2002.

Waldman, Murray, and Marjorie Lamb. *Dying for a Hamburger: Modern Meat Processing and the Epidemic of Alzheimer's Disease.* Toronto: McClelland & Stewart, 2004.

Williams, Elizabeth, and Mike Miller. "Transmissible Spongiform Encephalopathies in Non-domestic Animals: Origin, Transmission and Risk Factors." *OIE Scientific and Technical Review* 22, no. 1 (2003): 145-56.

Yam, Philip. *The Pathological Protein: Mad Cow, Chronic Wasting, and Other Deadly Prion Diseases.* New York: Copernicus Books, 2003.

To follow BSE's uncompleted odyssey, visit the Organic Consumers Association at http://www.organicconsumers.org/madcow.htm or the Web site for the Prepared Food and Meat Processing industry: http://www.meatprocess.com.

CHAPTER FIVE: RUSTS, BLIGHTS, AND THE INVADED LARDER

Ainsworth, G. *Introduction to the History of Plant Pathology.* Cambridge: Cambridge University Press, 1981.

Bandyopadhyay, Ranajit, and Richard Frederiksen. "Contemporary Global Movement of Emerging Plant Diseases." *Annals of New York Academy of Sciences* 894 (1999): 28-36.

Becker, Hank. "Fighting a Fungal Siege on Cacao Farms." *Agricultural Research,* November 1999. http://www.ars.usda.gov/is/AR/archive/nov99/cacao1199.htm.

Berry, Wendell. *The Art of the Common Place: The Agrarian Essays.* Washington, DC: Counterpoint Press, 2002.

Carefoot, G.L., and E.R. Sprott. *Famine on the Wind: Man's Battle Against Plant Disease.* Rand McNally, 1967.

Damsteegt, Vern. *New and Emerging Plant Viruses.* American Phytopathological Society, 1999. http://www.apsnet.org/online/feature/NewViruses/.

Doyle, Jack. *Altered Harvest: Agriculture, Genetics and the Fate of the World's Food Supply.* New York: Penguin, 1986.

Duncan, David. "Without a Genetic Fix, the Banana May Be History." *San Francisco Chronicle,* April 5, 2004.

Fry, William, and Stephen Goodwin. "Resurgence of the Irish Potato Famine Fungus." *BioScience* 47, no. 6 (June 1997): 363-71.

Harlan, Jack. *The Living Fields: Our Agricultural Heritage.* Cambridge: Cambridge University Press, 1995.

Heal, Geoffrey, et al. "Genetic Diversity and Interdependent Crop Choices in Agriculture." *Resource and Energy Economics* 26 (2004): 175-84.

Large, E.C. *The Advance of the Fungi.* New York: Dover, 1960.

Leonard, Kurt. "Black Stem Rust Biology and Threat to Wheat Growers." Presented to the Central Plant Board Meeting, Lexington, Kentucky, February 5-8, 2001. United States Department of Agriculture, Agricultural Research Service. http://www.ars.usda.gov/Main/docs.htm?docid=10755.

Mooney, Pat. "The Hidden Hot Zone: An Epidemic in Two Parts." *Journal of the Dag Hammarskjold Foundation* 2, no. 6 (August 1995).

Ordish, George. *The Constant Pest: A Short History of Pests and Their Control.* London: Peter Davies, 1976.

Ploetz, Randy. *The Most Important Disease of a Most Important Fruit.* American Phytopathological Society, 2001. http://www.apsnet.org/education/feature/banana/Top.html.

Rhoades, Robert. "The World's Food Supply at Risk." *National Geographic* 197, no. 4 (April 1991), 74-105.

Rosenzweig, Cynthia, et al. *Climate Change and U.S. Agriculture: The Impacts of Warming and Extreme Weather Events on Productivity, Plant Diseases, and Pests.* Boston: Centre for Health and the Global Environment, Harvard Medical School, 2000.

Windels, Carol. "Economic and Social Impacts of Fusarium Head Blight: Changing Farms and Rural Communities in the Northern Great Plains." *Phytopathology* 90, no. 1 (2000): 17-21.

CHAPTER SIX: RESURRECTING ANTHRAX

Alibek, Ken, with Stephen Handelman. *Biohazard: The Chilling True Story of the Largest Covert Biological Weapons Program in the World—Told from the Inside by the Man Who Ran It.* New York: Delta, 1999.

Broad, William, "In a Lonely Stand, a Scientist Takes On National Security Dogma." *New York Times,* June 29, 2004.

Bryden, John. *Deadly Allies: Canada's Secret War, 1937-1947.* Toronto: McClelland & Stewart, 1989.

Carroll, Michael. *Lab 257: The Disturbing Story of the Government's Secret Plum Island Germ Laboratory.* New York: William Morrow, 2004.

Cole, Leonard. *Clouds of Secrecy: The Army's Germ Warfare Tests over Populated Areas.* Totowa, NJ: Rowman & Littlefield, 1988.

———. *The Eleventh Plague: The Politics of Biological and Chemical Warfare.* New York: W.H. Freeman, 1997.

Foster, Don. "The Message in the Anthrax." *Vanity Fair,* October 2003.

Graysmith, Robert. *Amerithrax: The Hunt for the Anthrax Killer.* New York: Berkley Books, 2003.

Guillemin, Jeanne. "Biological Weapons and Secrecy." *FASEB Journal* (Federation of American Societies for Experimental Biology) 19 (2005): 1763-65.

Harris, Sheldon. *Factories of Death: Japanese Biological Warfare, 1932-1945, and the American Cover-up.* New York: Routledge, 2002.

Holmes, Chris. *Spores, Plagues and History: The Story of Anthrax.* Dallas: Durban House, 2003.

Kaufman, Arnold. "The Economic Impact of a Bioterrorist Attack: Are Prevention and Postattack Intervention Programs Justifiable?" *Emerging Infectious Diseases* 3, no. 2 (April-June 1997).

Matsumoto, Gary. *Vaccine A: The Covert Government Experiment That's Killing Our Soldiers and Why GI's Are Only the First Victims.* New York: Basic Books, 2004.

Nass, Meryl. "The Anthrax Vaccine Program: An Analysis of the CDC's Recommendations for Vaccine Use." *American Journal of Public Health* 92, no. 5 (May 2002). (See also Nass's excellent Web site: www.anthraxvaccine.org.)

Osterholm, Michael, with John Schwartz. *Living Terrors: What America Needs to Know to Survive the Coming Bioterrorist Catastrophe.* Toronto: Random House of Canada, 2001.

Preston, Richard. *The Demon in the Freezer.* New York: Random House, 2002.

Regis, Ed. *The Biology of Doom: The History of America's Secret Germ Warfare Project.* New York: Henry Holt, 1999.

Thompson, Marilyn. *The Killer Strain: Anthrax and a Government Exposed.* New York: HarperCollins, 2003.

CHAPTER SEVEN: MARINE INVADERS: CHOLERA'S CHILDREN

Awuonda, Moussa. "The Voices of the Dunga." Master's thesis no. 23, Swedish University of Agricultural Sciences, Uppsala, 2003.

Briggs, Charles, and Clara Mantini-Briggs. *Stories in the Time of Cholera: Racial Profiling During a Medical Nightmare.* Berkeley: University of California Press, 2003.

Carlton, James. "Environmental Impacts of Marine Exotics: An Action Bioscience.org Original Interview." American Institute of Biological Sciences, May 2004. http://www.actionbioscience.org/biodiversity/carlton.html.

Colwell, Rita. "A Global Thirst for Safe Water: The Case of Cholera." Abel Wolman Distinguished Lecture, National Academy of Sciences, January 25, 2002. http://sofia.usgs.gov/publications/lectures/safewater/.

Desowitz, Robert. *Federal Bodysnatchers and the New Guinea Virus: Tales of Parasites, People and Politics.* New York: W.W. Norton, 2002.

Dybas, Cheryl. "Harmful Algal Blooms: Biosensors Provide New Ways of Detecting and Monitoring Growing Threat in Coastal Waters." *BioScience* 53, no. 10 (October 2003): 918-23.

Goldschmidt, Tijs. *Darwin's Dreampond: Drama in Lake Victoria.* Cambridge, MA: MIT Press, 1998.

Harvell, Drew, et al. "Emerging Marine Diseases—Climate Links and Anthropogenic Factors." *Science* 285 (September 3, 1999).

―――. "The Rising Tide of Ocean Diseases: Unsolved Problems and Research Priorities." *Frontiers in Ecology and Environment* 2, no. 7 (2004): 375-82.

Jansen, Erik. "Out of a Lake." *Samudra,* September 1999.

Lee, Kelley, and Richard Dodgson. "Globalization and Cholera: Implications for Global Governance." *Global Governance* 6, no. 2 (April/June 2000): 213-47.

Low, Tim. *Ballast Invaders: The Problem and Response.* Invasive Species Council, Australia, September 2003.

McGrew, Roderick. *Russia and the Cholera, 1823-1832.* Madison: University of Wisconsin Press, 1965.

Meinesz, Alexander. *Killer Algae: The True Tale of a Biological Invasion.* Chicago: University of Chicago Press, 1999.

Molyneaux, Paul. "Disease: Shrimp Aquaculture's Biggest Problem." *APF Reporter* (Alicia Patterson Foundation) 21, no. 1 (2003).

Naylor, Rosamond, et al. "Aquaculture—A Gateway for Exotic Species." *Science* 294 (November 23, 2001): 1655-56.

Raaymakers, Steve. "The Ballast Water Problem: Global Ecological, Economic and Human Health Impacts." Paper presented at the RECSO/IMO Joint Seminar on Tanker Ballast Water Management & Technologies, Dubai, UAE, December 16-18, 2002.

Verity, Peter, et al. "Status, Trends and the Future of Marine Pelagic Ecosystems." *Environmental Conservation* 29, no. 2 (2002): 207-37.

Weiss, Kenneth. "Fish Farms Become Feedlots of the Sea." *Los Angeles Times,* December 9, 2002.

Yan, Tian. *A National Report on Harmful Algal Blooms in China.* North Pacific Marine Sciences Organization (PICES), Scientific Report no. 23, 2003. http://www.pices.int/publications/scientific_reports/Report23/default.aspx.

CHAPTER EIGHT: CLIMATE RIDERS

Becker, Amy. "Scientists Worry That Rift Valley Fever Could Reach US." *CIDRAP News* (Center for Infectious Disease Research and Policy, University of Minnesota), July 21, 2004.

Brownstein, John. "Effect of Climate Change on Lyme Disease Risk in North America." *EcoHealth* 2 (2005): 38-46.

Calisher, Charles. "Persistent Emergence of Dengue." *Emerging Infectious Diseases* 11, no. 5 (May 2005).

Epstein, Paul, and Evan Mills, eds. *Climate Change Futures: Health, Ecological and Economic Dimensions.* Boston: Centre for Health and the Global Environment, Harvard Medical School, November 2005.

Fagan, Brian. *The Little Ice Age: How Climate Made History, 1300 to 1850.* New York: Basic Books, 2000.

————. *The Long Summer: How Climate Changed Civilization.* New York: Basic Books, 2004.

Fenech, Adam, et al. "The Spread and Severity of the West Nile Virus in Ontario 2000-2004: Implications for Adaptation to Climate Change." Presented at Adapting to Climate Change in Canada, Montreal, May 4-7, 2005.

Food and Agriculture Organization (FAO). "Flood Victims in East Africa Hit by Rift Valley Fever Epidemic." News release, January 22, 1998. http://www.fao.org/news/1998/980105-e.htm.

Harvell, Drew, et al. "Climate Warming and Disease Risks for Terrestrial and Marine Biota." *Science* 21 (June 2002): 2158-62.

Junttila, Juha, et al. "Prevalence of *Borrelia burgdorferi* in *Ixodes ricinus* Ticks in Urban Recreational Areas of Helsinki." *Journal of Clinical Microbiology* 37, no. 5 (May 1999): 1361-65.

Kampen, Helge. "Substantial Rise in the Prevalence of Lyme Borreliosis Spirochetes in a Region of Western Germany over a 10-Year Period." *Applied and Environmental Microbiology,* March 2004, 1576-82.

Kilpatrick, Marm, et al. "West Nile Virus Risk Assessment and the Bridge Vector Paradigm." *Emerging Infectious Diseases* 11, no. 3 (March 2005).

Lindgren, Elisabet. "Climate and Tickborne Encephalitis." *Ecology and Society* 2, no. 1 (1998).

Schmidt, Kenneth, and Richard Ostfeld. "Biodiversity and the Dilution Effect in Disease Ecology." *Ecology* 82, no. 3 (2001): 609-19.

Smith, Patricia. "The Effects of Lyme Disease on Students, Schools and School Policy." *School Leader* (New Jersey School Boards Association), September/October 2004.

Spielman, Andrew, and Michael D'Antonio. *Mosquito: A Natural History of Our Most Persistent and Deadly Foe.* New York: Hyperion, 2001.

Sutherst, Robert. "Global Change and Human Vulnerability to Vector-Borne Diseases." *Clinical Microbiology Reviews* 17, no. 1 (January 2004): 136-73.

United States. General Accounting Office. *West Nile Virus Outbreak: Lessons for Public Health Preparedness.* Report to Congressional Requesters. Washington, DC: GAO, September 2000.

Weinberger, Miriam, et al. "West Nile Fever Outbreak, Israel, 2000: Epidemiologic Aspects." *Emerging Infectious Diseases* 7, no. 4 (July-August 2001).

Worster, Donald, ed. *The Ends of the Earth: Perspectives on Modern Environmental History.* Cambridge: Cambridge University Press, 1988.

For more information on Lyme disease, see the Web sites of the Canadian Lyme Disease Foundation, http://www.canlyme.org, and the International Lyme and Associated Diseases Society, http://www.ilads.org/. The leader in documenting the health consequences of climate change remains the Centre for Health and the Global Environment: http://chge.med.harvard.edu/.

CHAPTER NINE: THE GLOBAL HOSPITAL

Archer, Gordon. "*Staphylococcus aureus:* A Well-Armed Pathogen." *Clinical Infectious Diseases* 26 (1998): 1179-81.

Ayliffe, Graham, and Mary English. *Hospital Infection from Miasmas to MRSA.* Cambridge: Cambridge University Press, 2003.

Cameron, Peter. "The Plague Within: An Australian Doctor's Experience of SARS in Hong Kong." *Medical Journal of Australia* 178, no. 10 (2003): 512-13.

Canada. National Advisory Committee on SARS and Public Health. *Learning from SARS: Renewal of Public Health in Canada.* Ottawa: Health Canada, October 2003. http://www.phac-aspc.gc.ca/publicat/sars-sras/naylor/.

Dial, Sandra, et al. "Risk of *Clostridium difficile* Diarrhea Among Hospital Inpatients Prescribed Proton Pump Inhibitors: Cohort and Case-Control Studies." *Canadian Medical Association Journal* 171, no. 1 (July 6, 2004).

Eggertson, Laura, and Barbara Sibbald, "Hospitals Battling Outbreaks of *C. difficile.*" *Canadian Medical Association Journal* 171, no. 1 (July 6, 2004).

Illich, Ivan. "Medical Nemesis." *Lancet* 303, no. 7863 (1974): 918-21. Reprinted in *Journal of Epidemiology and Community Health* 57 (2003): 919-22.

Leung, P., and E. Ooi, eds. *SARS War: Combating the Disease.* London: World Scientific, 2003.

McCaughey, Betsy. *Unnecessary Deaths: The Human and Financial Costs of Hospital Infections.* New York: Committee to Reduce Infection Deaths, 2005. www.hospitalinfection.org /ridbooklet.pdf.

McGeer, Allison. "News in Antimicrobial Resistance: Documenting the Progress of Pathogens." *Infection Control and Hospital Epidemiology* 25, no. 2 (February 2004).

Ontario. SARS Commission. *SARS and Public Health in Ontario: Interim Report.* Archie Campbell, Commissioner. Toronto: the Commission, April 15, 2004. http://www.sarscommission.ca/report/Interim_Report.pdf.

Pennington, Hugh. "Don't Pick Your Nose." *London Review of Books,* December 15, 2005.

Risse, Guenter. *Mending Bodies, Saving Souls: A History of Hospitals.* Oxford: Oxford University Press, 1999.

Schull, M.J. "Sex, SARS and the Holy Grail." *Emergency Medicine Journal* 20 (2003): 400-401.

Shnayerson, Michael, and Mark Plotkin. *The Killers Within: The Deadly Rise of Drug-Resistant Bacteria.* New York: Little, Brown, 2002.

Tambyah, Paul. "The SARS Outbreak: How Many Reminders Do We Need." *Singapore Medical Journal* 44, no. 4 (2003): 165-67.

United Kingdom. Comptroller and Auditor General. *Improving Patient Care by Reducing the Risk of Hospital Acquired Infection: A Progress Report.* London: National Audit Office, July 2004.

————. National Clostridium difficile Standards Group. *Report to the Department of Health.* London: the Department, February 2003.

Vos, Margareet, and Henri Verbrugh. "MRSA: We Can Overcome, But Who Will Lead the Battle?" *Infection Control and Hospital Epidemiology* 26, no. 2 (February 2005): 117-21.

Wenzel, Richard, ed. *Prevention and Control of Nosocomial Infections.* New York: Lippincott Williams and Wilkins, 2003.

Other valuable MRSA sources include British Nursing News Online, http://www.bnn-online.co.uk/; the Consumers Union, http://www.consumers union.org/campaigns/stophospitalinfections/learn.html; MRSA Support, http://www.mrsasupport.co.uk/; and MRSA Watch, http://tahilla.type pad.com/mrsawatch/. In Canada, consult the Association to Defend Victims of Nosocomial Infections (ADVIN): http://www.advin.org.

EPILOGUE: THE NEXT PANDEMIC

An Analysis of the Potential Impact of the H5N1 Avian Flu Virus. Food Industry QRT Pandemic Analysis. August 2005. http://www.cidrap.umn.edu/cidrap/files/47/panbusplan.pdf.

Barry, John. *The Great Influenza.* New York: Penguin, 2004.

Butler, Declan. "The Flu Pandemic: Were We Ready?" *Nature* 435 (May 26, 2005): 400-402.

Camus, Albert. *The Plague.* Middlesex, UK: Penguin, 1968.

Cantor, Norman. *In the Wake of the Plague: The Black Death and the World It Made.* New York: Perennial, 2002.

Cooper, Sherry. "The Avian Flu Crisis: An Economic Update." Toronto: BMO Nesbitt Burns, March 13, 2006. http://www.cfr.org/publication/10129/bmo_nesbittburns.html.

Crosby, Alfred, Jr. *America's Forgotten Pandemic: The Influenza of 1918.* Cambridge: Cambridge University Press, 2003.

Dubos, Rene. *Mirage of Health: Utopias, Progress and Biological Change.* New York: Harper Colophon, 1979.

Girardet, Herbert. "Cities, People, Planet." Schumacher Lectures, Liverpool, U.K., April 8, 2000.

Glass, Thomas. "Understanding Public Response to Disasters." *Public Health Reports* 116, supplement 2 (2001).

Revill, Jo. *Everything You Need to Know About Bird Flu and What You Can Do to Prepare for It.* London: Rodale, 2005.

Welsh, Ian. *The Economics of a Pandemic.* Flu Wiki, 2005. www.fluwikie.com/index.php?n=Issues.EconomicWelsh.

A CANTICLE FOR LOCAL LIVING

Berry, Wendell. *The Way of Ignorance and Other Essays.* Emeryville, CA: Shoemaker and Hoard, 2005.

Cayley, David, ed. *The Rivers North of the Future: The Testament of Ivan Illich.* Toronto: Anansi, 2005.

Dalmazzone, Silvana. "Economic Factors Affecting Vulnerability to Biological Invasions." In *The Economic of Biological Invasions,* edited by C. Perrings, M. Williamson, and S. Dalmazzone, 17-30. Cheltenham, UK: Edward Elgar, 2000.

Epictetus. *The Art of Living.* Edited by Sharon Lebell. San Francisco: HarperCollins, 1995.

Hardin, Garrett. *Filters Against Folly.* New York: Viking, 1985.

MacIsaac, Hugh, et al. *Economic Impacts of Invasive Nonindigenous Species in Canada: A Case Study Approach.* Ottawa: Auditor General of Canada, May 2002.

Acknowledgments

B ird flu is not an aberration but a symptom of the global pandemonium caused by biological invasions. Rudolf Virchow, a wise 19th-
century pathologist and social crusader, believed that "science should
speak the language of the common people." To that end, I have tried to
synthesize the important research of many brilliant ecologists, historians, doctors, veterinarians, and biologists. The valuable works of Yvonne
Baskin, Wendell Berry, Corrie Brown, James T. Carlton, Eric Chivian,
Alfred Crosby, Jr., Peter Daszak, Andrew Dobson, Rene Dubos, Charles
Elton, Paul Epstein, Jack Harlan, Vaclav Kouba, William Leiss, Darrel
Rowledge, Daniel Simberloff, Rudolf Virchow, and the Consortium for
Conservation Medicine informed and guided this project.

Several scientists served as critical sources or readers over my
shoulder. Earl Brown, a virologist at the University of Ottawa, explained
the explosive relationship between bird flu and factory chicken densities for Chapter 1; Hugh MacIsaac, a specialist in invasive species at
the University of Windsor, offered suggestions and corrections for
Chapter 2; Drew Harvell, a marine ecologist at Cornell University,
outlined the frightful scale of plagues killing ocean life for Chapter 7;
Dr. Sandra Dial at the Montreal Chest Institute provided invaluable
insights on the *C. difficile* outbreak in Quebec for Chapter 9. The
indomitable David Schindler, Killam Memorial Professor of Ecology at

the University of Alberta and a former student of Charles Elton, gave corrections and feedback on various parts of the manuscript.

Crucial funding provided by the Canada Council and the Margaret Nockleberg Foundation kept this book alive.

Although this project was often derailed by illness, family emergencies, and Alberta's messy oil and gas politics, my friends Ed Struzik, Sid Marty, and Heather Pringle reminded me to laugh. They weren't always successful.

A patient team at Penguin Canada led by my editor, Diane Turbide, helped to produce this book with intelligence and flair. While Elizabeth McKay assembled the illustrations, Tracy Bordian efficiently kept the manuscript moving. A special thanks goes to the remarkable Barbara Czarnecki, Canada's best copy editor.

To my boys, Aidan, Keegan, and Torin, and my beautiful wife, Doreen, I owe an apology: This book simply invaded our lives and took too much time and energy to write. Thank you, as always, for your love and understanding.

I am responsible for any errors and encourage readers to send their comments to **andrew@andrewnikiforuk.com**.

Blue skies.

Index

Page numbers in italic indicate visuals.